Sustainable Agriculture in Brazil

NEW HORIZONS IN ENVIRONMENTAL ECONOMICS

General Editors: Wallace E. Oates, *Professor of Economics, University of Maryland, USA* and Henk Folmer, *Professor of Economics, Wageningen Agricultural University, The Netherlands and Professor of Environmental Economics, Tilburg University, The Netherlands*

This important series is designed to make a significant contribution to the development of the principles and practices of environmental economics. It includes both theoretical and empirical work. International in scope, it addresses issues of current and future concern in both East and West and in developed and developing countries.

The main purpose of the series is to create a forum for the publication of high quality work and to show how economic analysis can make a contribution to understanding and resolving the environmental problems confronting the world in the twenty-first century.

Recent titles in the series include:

The International Yearbook of Environmental and Resource Economics 1999/2000
Edited by Henk Folmer and Tom Tietenberg

Valuing Environmental Benefits
Selected Essays of Maureen Cropper
Maureen Cropper

Making the Environment Count
Selected Essays of Alan Randall
Alan Randall

Controlling Air Pollution in China
Risk Valuation and the Definition of Environmental Policy
Therese Feng

Sustainable Agriculture in Brazil
Economic Development and Deforestation
Jill L. Caviglia

The Political Economy of Environmental Taxes
Nicolas Wallart

Trade and the Environment
Selected Essays of Alistair M. Ulph
Alistair M. Ulph

Water Management in the 21st Century
The Allocation Imperative
Terence Richard Lee

Institutions, Transaction Costs and Environmental Policy
Institutional Reform for Water Resources
Ray Challen

Sustainable Agriculture in Brazil

Economic Development and Deforestation

Jill L. Caviglia
Salisbury State University, Salisbury, MD, USA

New Horizons in Environmental Economics

Edward Elgar
Cheltenham, UK • Northampton, MA, USA

Published by
Edward Elgar Publishing Limited
Glensanda House
Montpellier Parade
Cheltenham
Glos GL50 1UA
UK

Edward Elgar Publishing, Inc.
136 West Street
Suite 202
Northampton
Massachusetts 01060
USA

A catalogue record for this book
is available from the British Library

Library of Congress Cataloguing in Publication Data
Caviglia, Jill L., 1970-
     Sustainable agriculture in Brazil : economic development and
deforestation / Jill L. Caviglia.
     (New horizons in environmental economics)
     Includes bibliographical references (p.)
     1. Sustainable agriculture—Brazil—Ouro Preto (Rondônia) Case
studies. 2. Shifting cultivation—Brazil—Ouro Preto (Rondônia)
Case studies. 3. Deforestation—Brazil—Ouro Preto (Rondônia) Case
studies. I. Title. II. Series.
S475.B720933   1999
338.1'0981'11—dc21   99–22086
                                                             CIP

ISBN 1 84064 145 2

Printed and bound in Great Britain by Biddles Ltd, Guildford and King's Lynn

Contents

List of Tables

List of Figures

Acknowledgments

I would like to take this opportunity to thank the numerous individuals and institutions which have helped me complete this book. This research began as my doctoral dissertation and has been compiled into a book with the assistance of many individuals. I would not have been able to accomplish this amazing task without the support and encouragement of my family, friends and doctoral committee.

I would first like to thank my doctoral committee: Drs James Kahn, Jonathan Rubin, Matthew Murray and Virginia Dale, who all took the time and effort not only to read and edit my work but also to aid in the development of my theory and ideas. I would particularly like to thank James Kahn for opening my eyes to environmental economics and helping me see how exciting the discipline can be and for all of his patient advice in terms of this study and my future career; Jonathan Rubin for his friendship, guidance and constant push to keep me going during this research endeavor; Matthew Murray for his enthusiasm for economics and genuine concern for the graduate program; and Virginia Dale for her wealth of knowledge on tropical deforestation and pertinent related research in Brazil. I could not have completed this research without this combination of dedicated committee members. Additionally, I want to thank James Kahn for continuing his support and serving as my mentor after the completion of this project. His advice has been significant.

This research was supported by international grants and fellowships from the National Security Education Program, the Organization of American States, the Institute for the Study of World Politics and the McClure Fund Foundation. I would like to thank these institutions because without their financial support the data collection in Brazil would not have been possible.

In connection with the data collection, I would also like to thank the various individuals who assisted with the logistics of the research. First I would like to thank the faculty of the Environmental Sciences Research Division at the University of Manaus, Brazil. In particular, I would like to thank Dr Neliton Marques da Silva and his family, who so graciously took me in during my stay in Manaus. I also would like to thank Walmir de Jesus, a Rondônian farmer from the association APA. Walmir de Jesus aided in the interviewing process. Walmir was patient with my struggle with the Portuguese language, and gave me the opportunity to learn first hand about Rondônian farm life by arranging my stay at the homes of the farmers. He also became a great friend, and proved to be very knowledgeable in terms of sustainable methods of farming and many other pertinent issues. I look forward to working with him in the future. I also need to thank Christopher Brown and Denise Perpich for their assistance in Rondônia (in many different ways) and for their intellectual stimulation. Both of them come from different disciplines, and have helped me see the deforestation and farming situation in a different light. And finally, I would like to thank Marcos Pedlowski, who assisted me with making contacts in Rondônia. He also spent numerous hours speaking with me by phone and communicating with me through e-mail, helping to prepare me for the research project. I am indebted to Marcos for this assistance.

And last, I would like to thank my family, in particular my parents, for all of their patience and support throughout the years; and my sister, Christy, for her encouragement. I would not have been successful without their confidence in my ability.

Abstract

This book explores the relationship between the land use choices of small-scale farmers and the rate of deforestation in the Brazilian Amazon. The major source of deforestation in the Amazonian state of Rondônia is the use of slash-and-burn agriculture by small-scale farmers. The adoption of sustainable agriculture by these farmers could therefore reduce deforestation dramatically. Although sustainable agriculture was introduced to the area about ten years ago, only a few farmers have adopted it. Factors which may increase the rate of adoption of sustainable agriculture are investigated in order to identify policies that best address this issue.

The book begins by summarizing the history of the Amazon and the trends that have led to the development and deforestation of the region. The settlement of the Amazon is important to consider in the analysis of deforestation since migrants are responsible for a majority of the deforestation of the Amazon. Many of the migrants came from temperate regions of the country and were unaware of the best farming and ranching techniques for tropical soils. Many farmers are currently learning about sustainable agriculture, which is appropriate for tropical soils, through the assistance of extension groups. The role of sustainable agriculture is identified and defined.

The empirical analysis examines the adoption of sustainable agriculture by farmers in the settlement of Ouro Preto do Oeste, Rondônia based on a stratified random sample of 171 farmers and a sample of farmers who participate in the Association of Alternative Producers. A discrete-choice Heckman model is used to estimate the probability of adopting sustainable agriculture and the intensity of adoption (once the adoption decision is made). The analysis concludes that the adoption of sustainable agriculture is influenced greatly by farmer organizations and by the knowledge that sustainable agriculture is an alternative to slash-and-burn agriculture. Imperfect information appears to be the main deterrent to the adoption of sustainable agriculture. Local farm organizations provide both information about and experience with sustainable agriculture in a short period of time and therefore influence the adoption rate. In addition, farmer characteristics such as education level are found to play a role in the adoption decision. These results suggest that policies that provide information about sustainable agriculture and support local farm organizations can greatly increase the diffusion of sustainable agriculture in Ouro Preto do Oeste, Rondônia and similar settlements in the Amazon.

Preface

This book was developed from an interest in environmental economics, international issues and tropical deforestation. A collaboration between the University of Tennessee and the University of Amazonas, Manaus has not only made the collection of empirical data possible, but has also provided the opportunity for me to better understand the deforestation issues that impact Brazil and the local farmers who live in the Amazon. The time spent collecting the data for this study and living in the Amazon was an unforgettable experience. The hospitality of the faculty members at the University of Amazonas and the farmers in Ouro Preto do Oeste was remarkable and has helped to make this research successful. The study described in the pages that follow has allowed me to explore the many economic issues that impact tropical deforestation.

Chapter 1 is the introduction, containing the motivation and methods used in the study.

Chapter 2, 'Deforestation in the Brazilian Amazon', is a review of the Brazilian history that has led to the development of the Amazon and the beginning of a major part of the deforestation in the region.

Chapter 3, 'Sustainable Agriculture', provides a general definition of sustainable agriculture. This type of agriculture differs among tropical regions. The chapter continues with a specific definition of the type of sustainable agriculture to be used in this study in Ouro Preto do Oeste, Rondônia.

Chapter 4, 'Addressing Market Failure Which Has Led to Tropical Deforestation with Discrete Choice Models', provides the economic framework and theory that support policies which may reduce the deforestation rate in Brazil. The chapter continues with a literature review of economic studies on the adoption of new technologies in developing countries and relates this approach to technology adoption to the adoption of sustainable agriculture in the state of Rondônia, Brazil.

Chapter 5, 'The Survey', explains the survey design, the data collection process and presents some of the important data in tables.

Chapter 6, 'The Economic Model', is the data analysis section, in which regressions are estimated and interpreted according to economic theory. This chapter first presents the addition that this study can make to the economic literature, continues by presenting the economic theory to be used in the modeling, presents the regression results and makes policy recommendations which may reduce the deforestation rate based on these results.

Chapter 7 summarizes this study, makes policy implications as to the best ways that the adoption of sustainable agriculture may be increased and makes suggestions for further research.

Chapter 1

Introduction

1.1 TROPICAL DEFORESTATION

Tropical forests, which cover around 7.5 million km^2 of the humid tropics, are being destroyed at a progressively increasing rate (Myers 1994). Tropical deforestation is considered to be a major environmental crisis because of its international impacts on biodiversity loss and climate and because of its local impacts such as an increase in flood occurrence, the depletion of forest resources and soil erosion (Baer 1995, Myers 1994, Anderson 1993, Mahar 1989). Tropical deforestation is not a new phenomenon since it has been occurring since the beginning of human activity, but it is the acceleration in the rate of deforestation that has captured the concern of the world (Kahn 1995). Policies are urgently needed to slow deforestation, discourage unsustainable uses of the forest and to make sustainable uses profitable for the people who live in tropical forests (Fearnside 1990).

The determinants of participation in more sustainable uses of the forest are necessary to investigate if policies which promote alternative, more sustainable, uses of the forest are to be successful. One unsustainable use of tropical forests is slash-and-burn agriculture. Slash-and-burn agriculture is not the main source of deforestation in all parts of the world, but it is concentrated largely in the Amazon and Central America (Myers 1994). One way of reducing the rate of deforestation in tropical forests affected by slash-and-burn agriculture would be to decrease the number of farmers who practice this type of agriculture.

The factors that influence the use and adoption of an alternative to slash-and-burn agriculture are investigated using an empirical model of adoption. This model investigates the probability that small-scale farmers in the state of

Rondônia, Brazil, will adopt sustainable agriculture, and the extent of adoption once the adoption decision is made. The diffusion rate of this new farming technology is investigated according to the Mansfield model of technology adoption, in which the adoption rate of new technologies occurs according to a logistic shaped curve. What is interesting in Rondônia, and in many other developing countries, is that adoption has not occurred at a rapid rate even though the more sustainable methods of farming require less labor and can potentially increase profits (Holden 1993, Kebede *et al.* 1990, Sands 1986, Opare 1977). Policy implications are therefore made to address the low adoption rate of sustainable agriculture[1] in the settlement of Ouro Preto do Oeste, Rondônia.

1.2 RESEARCH FOCUS

This research will focus on the reduction of tropical deforestation through economic and welfare incentives which support the adoption of sustainable agriculture in the city of Ouro Preto do Oeste, Rondônia. Deforestation in this region of the Brazilian Amazon is mainly the result of small-scale farming. Deforestation rates can, therefore, be reduced by addressing the source of deforestation, the agricultural methods. Many of the farmers who live in the southern regions of the Amazon, including Rondônia, were enticed by the government programs to move to the area for free land and an offer of a better life. Responding to these economic incentives, immigrants from the south and northwest of the country migrated to the Amazon and brought with them agricultural methods traditional to the more temperate climates from which they came. Since farmers have migrated to the Amazon, these agricultural methods have been found to be inappropriate methods of farming in tropical soils since they strip the soils of their nutrients, rendering the land useless for agricultural purposes (Mahar and Schneider 1994, Frohn *et al.* 1990, Moran 1990).

Just as economic incentives have helped create a problem, economic incentives might be used to help remedy the current situation. Unfortunately, little is known about why sustainable agriculture is adopted by some farmers and not by others in the tropical forests of Rondônia. An analysis of the factors which increase the rate of adoption is therefore necessary in order to make the appropriate policy action suggestions.

Farmers' decisions and methods of agriculture are investigated from data collected in the city of Ouro Preto do Oeste, Rondônia, located in the southeastern border of the Brazilian Amazon. This area of the Amazon is an excellent location for this research for three reasons. First, Ouro Preto do

Oeste is located in the state of Rondônia, which is currently experiencing the highest rate of deforestation due to small-scale farming in the country. Second, Ouro Preto do Oeste is a relatively developed city in the Amazon making the study of the area and the settlers possible. And third, the city has recently established programs that support and encourage the use of sustainable agriculture. These programs include the dissemination of information about sustainable agriculture through workshops and the distribution of the seedlings used in intercropping.

It is expected that as more farmers adopt sustainable agriculture in Ouro Preto do Oeste, the deforestation rate will be reduced. The diffusion rate of this technology is, however, very low. The reasons why more farmers have not adopted this superior method of farming in terms of profit and sustainability are investigated. Implications are made as to the best way that the adoption of the more sustainable method of farming can be increased. The policy implications are, however, preliminary since this is the first step in a larger research objective. This research will be continued throughout the years as this relatively newly settled area of the Amazon is established and develops. Changes in agricultural techniques and deforestation rates will be recorded and analyzed as this research is extended.

The data used in this analysis were collected according to a time constraint and on a relatively small budget, and are therefore limited. At the same time, data remain sufficient for the analysis. In particular, the data allow the examination of the role that extension programs and farmer characterizations play in the adoption of sustainable agriculture in Ouro Preto do Oeste. The extension of the data from this pilot study, to broader research with fewer constraints, can lead to research results with broader policy implications.

1.3 THE ROLE OF SUSTAINABLE AGRICULTURE IN THE TROPICS

The alternative to slash-and-burn discussed in this chapter is known as sustainable agriculture. The terminology used in reference to sustainable agriculture is clarified here. The term 'intercropping' refers only to the alley cropping of perennials together, or with annual crops. 'Sustainable agriculture' refers to this intercropping practiced alone or in combination with other income sources that do not rely on the destruction of the rain forest, such as beekeeping and fish-raising. Therefore, the term 'sustainable agriculture' is used to reference both the use of intercropping and the income sources that do not rely on the destruction of the rain forest. A more detailed discussion of sustainable agriculture and intercropping can be found in Chapter 3.

Although sustainable agriculture can offer a solution to the increasing trends of deforestation, there is risk involved with adopting sustainable agriculture since developed markets have not been established to sell many of the native products of the forest which are harvested in intercropping plots. There is also risk involved because of the uncertainty as to how much income can be earned. Even though these deterrents exist, with the help of international grants some farmers choose to try this new method of farming in Rondônia.

It is important to point out that sustainable agriculture is not proposed here as a solution to the deforestation problem in Rondônia. Instead it is proposed as an alternative to slash-and-burn agriculture which can help to reduce the rate of deforestation by reducing the need for farmers to burn primary forest to plant crops. It is assumed that as more land is converted to sustainable agriculture, relative to slash-and-burn agriculture, that less land will be cut down and burned to support farm crops.

1.4 THE ECONOMIC MODEL

An economic model that illustrates the dichotomous choices between sustainable and slash-and-burn agriculture, and whether or not to join APA, will be used to observe the behavior patterns of farmers under different circumstances. A Heckman model of estimation is used to measure not only the probability of adoption of sustainable agriculture but also the intensity of use of sustainable agriculture once it is adopted.

Tropical deforestation has been addressed at length in the economic literature in terms of both the micro and macro levels. However, there is little research which has addressed deforestation in terms of its reduction through the adoption of a new technology. By combining the technology framework first introduced by Griliches (1957) with more advanced regression techniques to help solve market failure at the local level, this research provides a unique analysis to address deforestation reduction.

The following chapters in this study present the deforestation problem and the history of deforestation in Brazil and Rondônia state, explain the role that economics can play in the reduction of deforestation, analyze the situation in Ouro Preto do Oeste through economic modeling and conclude with policy implications.

ENDNOTES

1. Sustainable agriculture refers to agriculture that maintains the environmental and ecological integrity of the soil, water and land systems in the area while providing sufficient income to farmers through the intercropping of perennial and annual crops. The use of sustainable agriculture can reduce deforestation because, unlike slash-and-burn agriculture, it does not require the burning of forests in order to plant crops annually. Sustainable agriculture has been supported locally as well as globally because, not only is the rate at which the forest is destroyed reduced when these methods of farming are adopted, but also the income and utility of the farmers who adopt this method of agriculture are increased.

Chapter 2

Deforestation in the Brazilian Amazon

2.1 THE DEFORESTATION DILEMMA IN BRAZIL

Deforestation in Brazil presents a crisis of many dimensions involving international, national and local concerns. At the international level there is great concern about the total amount of depleted forest (Anderson 1993, 1990). At the national level the government is concerned with economic stability and social problems. And at the local level there are concerns about the impact of deforestation on the region's land, rivers and soils.

International, national and local concerns provide different incentives regarding the amount of deforestation that is desirable. Deforestation, however, impacts the local people, in particular local farm families, all the more because in many cases it is their land that is destroyed in the process. The tropical forests of Brazil have been settled by many small farm families and ranchers. The impacts of deforestation are, therefore, felt by these families who depend upon the land for their livelihood.

A brief discussion of the international, national and local issues follows. Although it is international concerns that have brought this crisis to the attention of the world, it is local impacts which are of greatest concern. The small-scale farmers who live in the Amazon have played an important role in deforestation and can also play an important role in preservation. It is therefore vital that the situation is addressed at this level.

2.1.1 International Concerns

The main reason for the worldwide attention on tropical deforestation in Brazil is related to its global impact and the large size of the forest. The Brazilian tropical rain forest is the largest contiguous tropical rain forest in the world (Anderson 1993, 1990). It is believed that large-scale tropical deforestation will contribute greatly to the greenhouse effect and a considerable loss of biodiversity. The Brazilian Amazon alone stores about 60 billion tons of carbon, equivalent to 8 per cent of the total carbon dioxide present in the atmosphere (Baer 1995). Burning the forest releases the carbon into the atmosphere, increasing the greenhouse effect. The magnitude of carbon dioxide increase resulting from deforestation in the Amazon is significant to cumulative global additions (Baer 1995, Pearce and Brown 1994, Schneider 1992a, Goldenberg 1992, Cockburn 1988). Carbon emissions from the Brazilian Amazon alone represented approximately 2 per cent of total world carbon emissions in 1990 (Goldenberg 1992).

Another international concern is that large-scale removal may bring about a reduction of biodiversity (Baer 1995, Pearce and Brown 1994, National Academy Press 1992, Anderson 1990, Mahar 1989). Tropical forests are the most species-rich ecosystems in the world, and the Brazilian rain forest is no exception (Hartshorn 1992). The number of species worldwide is not known, but it is believed that tropical forests worldwide contain about one half and possibly up to 90 per cent (Miller *et al.* 1993) of the world's estimated 5 to 10 million species in just 7 per cent of the earth's surface (Pearce and Brown 1994, Southgate 1994, Mahar 1989). For example, the Amazon rainforests support 30,000 species of plant life, whereas South American temperate forests support 10,000 species of plant. In contrast to these numbers, North American forests normally contain only 10 to 15 species of plant (Mahar 1989). The quantity and variety of fish, bird and insect life are also unmatched. In the Brazilian rain forest there are over 2,000 known fish species compared to 250 species in the Mississippi River system and the 200 species found in all of Europe (Mahar 1989). The destruction of the rain forest affects populations of these species. It is feared that if the Amazon rain forest is destroyed many of these species will become extinct. Many of the species that have been lost have not been identified. Their inherent nature and aesthetic value along with potential agricultural and pharmaceutical values vanish with them (National Academy Press 1992).

Biodiversity within the tropical rain forest and many similar ecosystems is essential to the balance and survival among organisms. Each species in an ecological system is likely to have a worth to the community far beyond its intrinsic value (Tietenberg 1988). Species contribute balance and stability to their ecological communities by providing food sources and nutrients, or by holding the population of another species in check, thus increasing the likelihood of survival for each species in the community. Pollination systems, seed dissemination and plant vegetation usually involve highly co-evolved species in tropical rain forests (Hartshorn 1992). For example, together several species of the Heliconia flower provide hummingbirds with year-round nectar. Responding to the supply of nectar, hummingbirds that live in tropical forests raise their young during the period when there is maximum nectar production (Hartshorn 1992). This interdependence among species makes each individual species more resilient and productive when the ecosystem is intact, but more vulnerable when one part of the system is adversely affected.

2.1.2 National Concerns

At the national level, the Brazilian government is concerned with an increasing trend of poverty and over-population in the southern and northeastern regions of the country, especially in the populous southern urban areas. A colonization program of the Amazon Basin initiated in the late 1960s was intended to ameliorate the situation, but instead resulted in an increase in deforestation, many unsuccessful farm establishments and the beginning of slash-and-burn agriculture in the Amazon region by small-scale farmers. The colonization program known as 'Operation Amazonia' is discussed in detail in the following sections.

2.1.3 Local Concerns

At the local level, depletion of forest resources, soil erosion, increasing recurrence of floods and the loss of biodiversity have occurred. Up to now, no human activity has affected more land in Amazonia or been more critical to human well-being than slash-and-burn agriculture (Myers 1994, Uhl *et al.* 1989). The essential characteristic of such a cultivation process is that an area of forest is cut and burned to grow subsistence crops (Uhl *et al.* 1989). The land is typically cultivated until it can no longer support the harvest, turned over to pasture and later abandoned. In addition, the economic failure of many slash-and-burn farmers has led first to rapid growth and then to impoverishment of rain forest cities as failed farmers migrate to rain forest urban areas such as Manaus, Belem and Porto Velho.

2.2 REFORESTATION IN TROPICAL SOILS

Tropical deforestation is a critical problem because regeneration of the forest is often difficult and, at times, impossible. Unlike non-tropical forests, the soil in tropical forests is commonly devoid of many nutrients. Most nutrients are stored in the very thin layer of topsoil and the biomass of living organisms. The cycling of nutrients takes place through the litter fall and root uptake in tropical forests (Anderson 1990, Serrão and Toledo 1990). In addition, microorganisms in the litter help make nutrients available to trees. Removal of the trees depletes essential nutrients from the ecosystem while escalating erosion, soil compaction and weed infestation, and disturbs the equilibrium created between nutrient storage and forest cover (Anderson 1993). Agriculture tends to rapidly exhaust the natural fertility by interrupting the litter fall and root uptake cycle. Eventually, five to ten years after cutting down the forest, the soil becomes void of most fertility and is abandoned.

Reforestation is difficult and costly. The long-term recovery after deforestation of tropical forests is what makes tropical deforestation so critical. Natural vegetation can recover after deforestation in some circumstances, but the loss of biodiversity of the primary forest often cannot be reversed (Uhl *et al.* 1989). Over-intensive use of deforested sites in tropical rain forests can lead to permanent degradation of regional ecosystems (Anderson 1990, Serrão and Toledo 1990, Uhl *et al.* 1988).

An example of the reforestation problem with tropical forests can be found in eastern Amazonia in the state of Pará. An area of 35 000 km² was abandoned after a short-lived agricultural settlement. This area was monitored for regeneration, and after 50 years there is still little vegetation beyond scrub and brush growth (Myers 1994). Areas similar to this site that still have scant forest regrowth can be found in Columbia (Battjees 1988), Ecuador (Gentry 1989) and Peru (Dourojeanni 1988). There have also been cases where reforestation in poor tropical soils occurred. Buschbacher *et al.* (1988) found that some agricultural pastures in Pará, Brazil, were able to recover 25 per cent of the biomass of mature rainforest within eight years. These pastures were not used by farmers for long periods of time. Instead, the pastures were immediately abandoned after burning because of poor seed germination or for other management considerations. As the number of years that the land was used for agricultural purposes increased, Buschbacher *et al.* found that the degree of regeneration was reduced considerably. Nepstad *et al.* (1990) conclude in their study on tropical forest regeneration that it may take centuries for forest regrowth to take place on abandoned pastures that have histories of heavy use.

2.3 DEFORESTATION IN THE BRAZILIAN TROPICAL RAIN FOREST

Brazil contains over one half of South and Central America's tropical forests

Sustainable Agriculture in Brazil

(well over one quarter of the rain forests worldwide) and deforestation is increasing at an alarming rate. In 1989 it was estimated that 30 000 km² were deforested in Brazil (1.4 per cent of the total tropical forest) compared to 23 000 km² in 1979 (Myers 1994). The total deforestation estimated to have occurred up until 1989 is 660 000 km². In comparison, Indonesia had the second highest level of deforestation for 1989 at 12 000 km² (Table 2.1). The rate of deforestation in Brazil is below the worldwide average of 1.4 per cent for that year, and the reason the percentage of forest burned is this low in Brazil is due to the fact that the country contains the world's largest expanse of intact forest in the world (Anderson 1993). When comparing different rates of deforestation, it is important to note that the sources of tropical deforestation differ among regions of the world (Table 2.2).

Table 2.1 Deforestation Statistics Tropical Moist Forests: 1989

Country	Area (km²)	Original Extent of Forest Cover (km²)	Extent of Forest Cover 1989 (km²)	Amount of Deforestation in 1989 (km²)	(%)
Brazil	8 511 960	2 860 000	2 200 000	30 000	1.4
Columbia	1 138 891	700 000	278 500	6500	2.3
India	3 287 000	160 000	165 000	4300	2.6
Indonesia	1 919 300	1 220 000	860 000	12 000	1.4
Malaysia	329 079	305 000	157 000	4800	3.1
Mexico	1 967 180	400 000	166 000	7000	4.2
Myanmar	696 500	500 000	245 000	8500	3.5
Nigeria	924 000	72 000	28 000	4000	14.3
Thailand	513 517	435 000	74 000	6000	8.1
Zaire	2 344 886	1 245 000	1 000 000	4200	0.4

Source: N. Myers (1994), 'Tropical Deforestation: Rates and Patterns', p. 30 in *The Causes of Tropical Deforestation: The economic and statistical analysis of factors giving rise to the loss of tropical forests* (Brown and Pearce, eds), University College London Press, London. This table includes the ten countries which have had the highest amount of deforestation in 1989.

2.4 THE EXTENT OF DEFORESTATION IN THE BRAZILIAN STATES

On a percentage basis, deforestation in Brazil has been the highest in the states of Rondônia and Mato Grosso, followed by Maranhão, Acre and northern Goiás. The northern states of Roraima and Amapá as well as the northern portions of Pará and Amazonas have, so far, experienced little deforestation within their borders (Table 2.3).

Between 1988 and 1993 deforestation in the state of Rondônia, the state with the highest rate of deforestation, has increased by 10 050 km² to 68 050 km² or to 28 per cent of the total state land area (Pedlowski *et al.* 1997). The total deforestation, with respect to the land area of the state, increased by 4.3 per cent in five years.

Deforestation in the Amazon core is still limited. Most of the clearing has been concentrated largely in the Amazon periphery, the east-southeast-south-southwest perimeter. There exists a large area of the Amazon which is still in its original state, suggesting that policies that advance the conservation of the rain forest should concentrate on keeping this extensive area intact.

Little deforestation occurred in Brazil's Amazon before the 1970s. As of 1975, a total of only 3 million hectares or about 0.6 per cent of the Brazilian Amazon had been cleared. Between 1975 and 1987 the rate of deforestation steadily accelerated, increasing to 12.5 million hectares by 1980, and the rate is constantly increasing (Moran 1993). The recent occupation and intensified

Table 2.2 Worldwide Tropical Deforestation: 1991

Source of Deforestation	Area Deforested (km²)	Location of Deforestation
Slash-and-burn Agriculture	87 000 km²	Almost entirely confined to Central America and the Amazon region
Cattle Ranching	15 000 km²	
Roads and Mining	10 000 km²	
Commercial Logging	45 000 km²	2/3 of this total is from Southeast Asia

Source: N. Myers (1994), 'Tropical Deforestation: Rates and Patterns', p. 32 in *The Causes of Tropical Deforestation: The economic and statistical analysis of factors giving rise to the loss of tropical forests*, (Brown and Pearce, eds.), University College London Press, London.

Sustainable Agriculture in Brazil

Table 2.3 Deforestation in Brazil by State

State	Area in Legal Amazonia (km²)	Area Cleared Through 1988 (km²)	Area Cleared Through 1990 (km²)	% of Area Cleared Through 1988	% of Area Cleared Through 1990
Acre	152 589	19 500	21 332	12.8	14.0
Amapá	140 276	572	6417	4.1	4.57
Amazonas	1 567 125	105 790	204 519	6.8	13.1
Maranhão	257 451	50 670	57 106	19.7	22.2
Mato Grosso	881 001	208 000	296 100	23.6	33.6
Pará	1 248 042	120 000	181 154	9.6	14.5
Rondônia	243 044	58 000	65 534	23.7	27.0
Roraima	230 104	3270	5341	1.4	2.3
Goiás	285 793	33 120	1715	11.6	12.9
Total	**5 005 425**	**598 922**	**839 218**	**12.0**	**16.8**

Sources: 1988 statistics from A. B. Anderson (1993), 'Deforestation in Amazonia: Dynamics, Causes and Alternatives', p. 1 in *The Earthscan Reader in Tropical Forestry* (Simon Rietbergen, ed.), Earthscan Publishers, London. The table includes all nine states included in legal Amazonia borders.
1990 statistics based on Table 2 p. 109 in J. Goldenberg, 'Current Policies Aimed at Attaining a Model of Sustainable Development in Brazil', *Journal of Environment and Development* 1 (1), 105–115 (Summer 1992).

use of the Amazon forest has been concentrated in the southern and eastern areas of the Basin. The rain forest is being destroyed by cattle ranchers responding to government initiatives and subsidies, migrants engulfed in poverty and fleeing over-populated cities, loggers interested in precious hardwood, mining enterprises, land speculators and hydroelectric projects (Baer 1995). Of these sources of deforestation, agriculture has made the largest single impact (Southgate 1994, Baer 1995, Myers 1994, Mahar 1989).

2.5 OURO PRETO DO OESTE, RONDÔNIA

Operation Amazonia,[1] migration and government incentives have had profound impacts on deforestation in Brazil and, in particular, on the settlement of Ouro Preto do Oeste, which lies in the state of Rondônia, by introducing small-scale agriculture to the area. The analysis performed in this study investigates how the deforestation rate may be reduced as agricultural methods are changed in Ouro Preto do Oeste. This small Amazonian city has been impacted by governmental settlement programs administered in the 1960s through the 1980s. Ouro Preto do Oeste is situated along BR-364, the 1500 km Cuiabá–Porto Velho highway. A map of the state can be found in Figure 2.1,[2] where the city of Ouro Preto do Oeste and the six municipals or regions of the city are outlined.

At the turn of the century this part of eastern Amazonia was a rich rubber-producing area. Large-scale settlement began during the late 19th century, which brought an estimated 103 000 immigrant workers to Rondônia by 1912 (Browder 1994). Following the decline in the Amazon rubber trade, economic activity stagnated until the 1940s. In the 1950s the area became an important source of tin and gold; however, the area was virtually inaccessible by land. In 1965 Rondônia's population numbered only 70 000 and the forest, which covered 80 per cent of the state, was still intact.

The highway BR-364 is the only major paved road in Rondônia. It runs

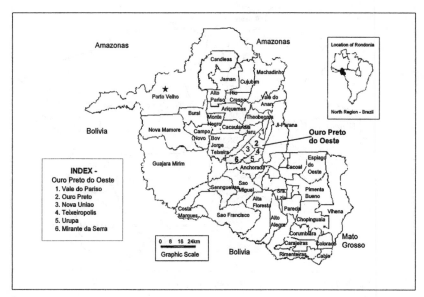

Figure 2.1 The State of Rondônia

from the southernmost city of Cuiabá and continues to the northernmost city, Porto Velho, the capital of the state (see Figure 2.2). With the completion of BR-364 in 1968, migration increased tenfold and clearing of the forest was up by 3 per cent within the decade. By 1988 this percentage was up to 23.7 per cent (Mahar 1989). The regional makeup of the population also changed. Previously the vast majority of Rondônia's settlers were from the north and northeast. The new surge of migration brought small-scale farmers in large part from the southern state of Paraná and also from the states of Mato Grosso, Minas Gerais, Espírito Santo and São Paulo (Mahar and Schneider 1994, Pedlowski and Dale 1992). The government programs, called 'Operation Amazonia', were responsible for much of the migration to this area and, therefore, much of the deforestation. A discussion of these government programs and the impact that they have made on the tropical forest follows.

Figure 2.2 Map of the Amazonian Highway System in Brazil

2.6 THE COSTS OF DEFORESTATION

Specific numbers as to the costs and benefits of deforestation in Brazil, or any other country, are difficult to estimate since these values can differ among the various levels of occurrence – international, national and local – and since the estimation of many public goods is necessary. One interesting cost is the damage due to the carbon dioxide release due to deforestation. Pearce and Brown (1994) report an estimation of international costs in terms of carbon release. It is estimated that the conversion of primary forest into agriculture would give rise to carbon release damage of about $4000 – $4400 per hectare.[3] The estimates allow for carbon fixation in the subsequent land use. In comparison, Schneider (1992b) reports an upper bound for land value in Rondônia of $300 per hectare, while the data collected for this study find an upper bound for land value of $440 per hectare. This substantial difference between the international damage value and the price of land represents a market failure on the global level.[4] Market failures, on all levels, are discussed in detail in the following chapter.

In a similar study Grimes *et al.* (1994) calculate the value of a hectare of primary forest in the Ecuadorian Amazon based on the potential extraction of non-timber forest products. Sustainable agriculture plots are determined to have the highest net present value. The authors find that the net present value of one hectare of intercropping ranged from $1257 to $2939, one hectare of agriculture was valued at less than $500 and that one hectare of land devoted to cattle raising ranged from $57 to $287.

2.7 CAUSES, SOURCES AND FORCES OF DEFORESTATION

It is important to make distinctions between *sources* and *causes* of deforestation because of the different implications of these terms. Panayotou and Sungsuwan (1994) define the *causes* of deforestation as population pressure, migration and social pressures and the *sources* of deforestation as logging, slash-and-burn agriculture and the others mentioned above. Pearce and Brown (1994) add to this definition fundamental forces which give rise to deforestation. These forces include the competition between humans and non-humans for land, and economic market failures.

In the Amazonian state of Rondônia, the prime cause of deforestation is the government incentives to move and live in the Amazon, while the sources are slash-and-burn agriculture, and the forces of deforestation are the economic failures.   The small-scale farmers are, therefore, not solely responsible for the deforestation in Rondônia.   Instead the farmers are recognized as poor individuals with few alternatives.   They are motivated to provide a subsistence living for themselves and their families with the available resources.

2.8 REDUCING DEFORESTATION IN RONDÔNIA

In the Brazilian Amazon general tax policies, special tax incentives, the rules of land allocation and poor government planning all helped to accelerate deforestation[5] in combination with population growth, and the city of Ouro Preto do Oeste, Rondônia, was no exception (Binswanger 1991).   If all deforestation is to be halted in Rondônia, the farmers, cattle ranchers, miners, loggers and families would have to be moved out of the area.   Halting all development in Rondônia is an unrealistic and undesirable goal. More feasible approaches concentrate on decreasing the rate of deforestation toward a steady-state zero rate of deforestation some time in the near future, while taking into account ecological, social and economic factors (Moran 1990). Sustainable agriculture can achieve these goals by allowing farmers and families to remain on their land longer and help to improve their welfare (Dale *et al.* 1994, Dale *et al.* 1993, Browder 1992).   This agricultural technique is discussed in detail in Chapter 3.

ENDNOTES

1. Operation Amazonia was a government-initiated program launched in 1964 to establish settlements in the undeveloped Amazon region. This program was initiated in response to many social and political problems which existed in the populated regions of the northeast and southeast. See Appendix A for a detailed account of Operation Amazonia.
2. Source: Secretaria de Estado de Desenvolvimento Ambiental (SEDAM) 1996. 'Mapa Politico e Administrativo de Rondônia'.
3. This estimate is based on a study by Fankhauser (1993), which suggests a central value of $20 of damage for every ton of carbon emitted.
4. In 1996 dollars the carbon damage from carbon release per hectare is $4338–$5432 and the Schneider land value is $322. After correcting for inflation, the results are the same. The damage of carbon emitted is far greater than the value of the land, reflected in the price of land.
5. The incentives created by tax policy, land allocation and government planning which led to the development and destruction of the Brazilian Amazon were initiated by Operation Amazonia. See Appendix A for a detailed account of this government program.

Chapter 3

Sustainable Agriculture

3.1 AGRICULTURAL PRACTICES IN THE BRAZILIAN AMAZON

The most common agricultural practice used in the Brazilian Amazon is slash-and-burn agriculture. It has been observed that subsequent to burning, forest agriculture can be achieved for an average of ten years (Browder 1994, Dale *et al.* 1994, Pedlowski and Dale 1992, Dale and Pedlowski 1992, Frohn *et al.* 1990, Mahar 1989, Goodland 1980). A common land use pattern is that farmers use the land for annuals crops for the first two to three years, and then convert the land to pasture and use it for this purpose until the land can no longer support grass (Dale *et al.* 1994, Pedlowski and Dale 1992, Dale and Pedlowski 1992, Fearnside 1990). After the soil is depleted of its nutrients, the farmer typically moves on to a new plot of land, resulting in the slashing and burning of more of the forest.

    The small-scale farmer is an important element of the deforestation crisis in Brazil because slash-and-burn agriculture is responsible for a major part of the deforestation (Myers 1994). One way of reducing deforestation would be to move these farm families, who use slash-and-burn agriculture, out of the rain forest. This potential way of reducing the deforestation rate is, however, unrealistic and undesirable. The role that more sustainable methods of agriculture can play in the preservation of the tropical forest is, therefore, crucial to addressing the deforestation dilemma. Altering farming techniques could drastically reduce the deforestation rate if adopted by a majority of the farmers. This chapter continues by addressing slash-and-burn agriculture, sustainable agriculture and the impacts that these agricultural methods make

17

on crop production.

3.1.1 Land Use Trends

Farmers in the Brazilian Amazon typically follow the land use pattern of burning the forest or older crops, planting annuals and then converting the land to pasture. The land is typically converted to pasture when the soil decreases in quality and can no longer support crops. A study that measured the rate of decreasing soil quality in the Brazilian tropics is by Fearnside (1990). Measurements of soil production over a full annual cycle in Ouro Preto do Oeste, Rondônia, show that a 12-year-old pasture produces at about half the rate of a three-year-old one. Fearnside found that yields decline due to the invasion of inedible weeds, soil compaction and decreasing levels of available phosphorous of the soils (Fearnside 1990, Hecht 1981, 1983).

The tropical soils of the Brazilian Amazon are acidic (have a very low pH level), have high aluminum saturation, low exchangeable bases and low nutrient content, adding up to low fertility (Villachia *et al.* 1990). Clearing and burning the tropical forests produces a temporary but substantial increase in soil fertility. Dramatic increases in potassium, calcium, magnesium, phosphorous and pH, as well as the reduction in aluminum saturation, occur due to this process (Serrão and Toledo 1990). The lack of phosphorous in tropical soils has been found to be the most limiting nutrient to agricultural systems followed by the lack of nitrogen (Serrão and Toledo 1990, Fearnside 1990). After the initial surge of nutrients being provided to the soil by burning, the nutrient base will become exhausted or limited after only three to five years (Serrão and Toledo 1990). The use of leguminous woody species (perennial crops) can therefore add fixed nitrogen to these soils if used and managed by a sustainable agricultural system (Nair 1992, Fearnside 1990).

Rice, beans and corn represent 58 per cent of the total area crop in Brazil, while three tree crops – banana, cocoa and coffee – comprise 37 per cent. The use of trees, native to the area, is not common practice. The second important trend of farmers in Rondônia has been the extensive conversion from crop cultivation to pasture formation and beef cattle production. Seeded pasture grass has become the dominant non-forest land use in Rondônia, increasing from 9.8 per cent of the total area in 1980 to 25.6 per cent in 1985 (Browder 1994).

3.2 DEFINITION OF SUSTAINABLE AGRICULTURE

This study explores the use of an alternative method of agriculture, known as sustainable agriculture, to potentially reduce deforestation in the parts of the Brazilian Amazon where slash-and-burn agriculture is practiced. Sustainable agriculture provides the soil with nutrients by maintaining the forest cover and leaf regenerating systems (in particular nitrogen), thus expanding the time farmers can cultivate the land almost indefinitely (Nair 1992). Sustainable agriculture has the potential to increase both the welfare of society by reducing the need to cut down the forest to plant crops and the welfare of farmers by reducing the labor input necessary for crop production and increasing income.

Sustainable agriculture is defined as agriculture that maintains the environmental and ecological integrity of the soil, water and land systems in the area while providing sufficient income to farmers through the intercropping of different perennials in combination with other perennial trees and/or annual crops. In Brazil, this type of agriculture refers to the farming that maintains the quality and nutrients of the soil, permitting long-term use of each lot of land (Prinsley 1992). The extremely site-specific nature of sustainable agriculture makes it difficult to develop precise recommendations for different areas around the world, but some key concepts are well established. Sustainable agriculture in the Amazon involves intercropping, or alley cropping, which has several characteristics:

1. It combines the production of multiple farm crops with protection of the resource base.
2. It emphasizes the use of indigenous multipurpose trees and shrubs helping maintain the natural balance of ecosystems.
3. It is a well-planned system with positive social, economic and ecological effects.
4. It is structurally and functionally more complex than monoculture (Nair 1992, Luo and Han 1990).

Amazon farming will probably never be sustainable in the sense that it can fully replace all of the natural resources that it depletes. However, as Almeida and Campari (1995) define it, Amazonian farming can be considered sustainable when the ratio of private benefits and social costs to the environment rises (i.e., when the private benefits increase at a more rapid rate than social costs). Almeida and Campari continue by adding that sustainability can be increased in this setting by increasing the private cost of deforesting land either by internalizing the social cost of deforestation or by decreasing the cost of alternative agricultural systems such as sustainable agriculture.

Sustainable agriculture as used by small-scale farmers in Ouro Preto do Oeste, Rondônia, includes the intercropping of tree crops such as cupuaçu, pupunha, açai, acerola, guarana, cacao, banana, orange, lime, coffee and trees that have precious woods such as mahogany, Brazil nut and rubber, in combination with annuals such as rice, corn, beans, sugar cane and manioc. The forest over-story (tallest tree layer) is generally dominated by rubber, mango, Brazil nut, mahogany, pupunha and other trees with valuable woods. The under-stories (lower-standing trees) are composed of shade-tolerant trees such as coffee, cacao, guarana and cupuaçú, intercropped with annuals. In Ouro Preto do Oeste these systems are often used by small-scale farmers in combination with other sustainable non-agricultural uses of the forest such as honey and fish harvesting.[1]

Tree crops protect the soil from degradation caused by leaching, erosion and compaction, and they frequently have a higher tolerance to soil acidity and aluminum toxicity than annual crops (Browder 1994). As is consistent with the tropical soils of the Brazilian Amazon, most of the 17 major soil groups found in Rondônia are distrophic, high in acidity and aluminum content, interspersed with patches of well-drained nonhydromorphic, red-yellow eutrophic podzols (Browder 1994). Tree crops also make relatively low demands on soil nutrients, in comparison to annual crops, because of efficient nutrient recycling. Important to the success of sustainable agriculture systems is the choice of trees that can be harvested for their fruits, wood or nuts and sold in the market (Uhl *et al.* 1989).

Dale *et al.* (1993) found that farmers in Rondônia who used innovative management systems for their farms (such as growing predominantly perennial crops and intercropping trees such as rubber, cocoa, palm oil, coconuts or Brazil nuts) made a better livelihood than those who used traditional methods. These farmers did not have pastures or income from cattle but averaged more than twice the income in 1990 of traditional farmers. They also had a much higher probability of staying on their original plot of land.

Although sustainable agriculture has shown some promising preliminary increases in soil quality and/or soil integrity (Ruddell 1995, Adriance 1995, Nair 1992 , Dale *et al.* 1993), sustainable agriculture cannot be considered a cure-all for tropical farming problems. One problem with sustainable agriculture is that plans cannot be transferred from one region of the world to another, but rather must be uniquely designed for each use and soil type. Another drawback is that most of the intercropping plots are fairly new. There is no specific evidence as yet that sustainable agriculture systems can be cultivated infinitely, although this may be possible. Compounding the problem of the lack of scientific evidence to support the long-term use of the soils in these sustainable systems is that many tropical soils differ so greatly from plot to plot, making numerous studies necessary. Currently studies are being done on the soil integrity of sustainable agricultural plots in Rondônia,[2] but since the plots are less than five years old, conclusive results are not expected for at least ten years. As Fearnside (1990) has pointed out, research in agroforestry and the effects of sustainable agriculture in the tropics is still in its infancy.

3.3 THE USE OF SUSTAINABLE AGRICULTURE AROUND THE WORLD

The use of sustainable agriculture by small-scale farmers has been increasing around the world. In Peru, a program that supports sustainable agriculture and the use of data collection as a tool for learning began in 1990 with the support of the grassroots organization World Neighbors[3] (Ruddell 1995). After only one planting season, yields and incomes increased for the participating farmers. In Guatemala, a group of dedicated farmers under Marcos Orozoco, a former Minister of Agriculture, has promoted sustainable agricultural methods. The results have been a reduction in slash-and-burn agriculture throughout the country, increased yields and the increased use of natural fertilizers (Adriance 1995). In Honduras, the group Familia y Medio Ambiente has found that the sustainable agricultural practice of intercropping has increased yields by using flowering plants as an aid in pest control.

Similar success can occur in Ouro Preto do Oeste if one can determine the best way to increase the adoption rate of sustainable agriculture. One way of motivating farmers to switch to this type of agriculture in Brazil is first to determine the factors that the farmers respond to best – such as subsidy and loan programs, education programs or information about this type of agriculture – and then to work with the local people to determine the best approach to the problem. Educational and outreach programs have been administered in Ouro Preto do Oeste, Rondônia. The educational programs consist of annual union meetings, which inform and assist farmers in the use of sustainable agriculture. While the outreach programs were initiated in the 1980s by a local NGO (non-governmental organization), IPHAE (Instituto de Pre-História, Agricultura, e Ecologia), the adoption rate has been low. The influence that these programs have on the adoption of sustainable agriculture in Ouro Preto do Oeste is investigated in Chapter 6. These programs are treated as independent variables in the regression analysis to determine their effects on the adoption and extent of the use of sustainable agriculture.

The local support of small farmers is necessary since financial support has been difficult to obtain from banks and government organizations in Central and South America for the peasant farmer. Many government programs that support improvements in agriculture have been designed primarily for the economies of scale of large farms with access to the financial capital to buy machinery, chemical fertilizers and pesticides. Therefore groups like World Neighbors and similar non-government organizations that provide access to capital have made a significant impact on the adoption of sustainable agriculture in developing nations. These groups have found that programs which involve local farmers have had great success.

3.4 FINANCING OF SUSTAINABLE AGRICULTURE PROGRAMS

One way of financing sustainable agriculture in a self-sufficient manner is through micro-loans to the farmers. Banks providing low-interest loans to small enterprises have become successful in the developing countries of Latin America, Africa and Asia. The Grameen Bank is perhaps the most successful and best publicized. This bank, an institution that pioneered lending to the landless poor in Asia's poorest country, Bangladesh, was founded by Muhammad Yunus, an economics professor at Bangladesh's Chittagong University (Stackhouse 1994). Since 1976, Grameen's 1.8 million borrowers have deposited about $50 million with the bank. Grameen customers, whose only collateral is the shirts on their backs, have borrowed $844 million and have repaid 98 per cent of their loans. This repayment percentage is above the performance of most commercial banks, which average less than 93 per cent, and above the performance of Japan's biggest banks, loaning to major industrial clients, which have a default rate of 3 per cent (Fragasso 1995). Grameen's record has been so successful that governments from Malaysia and several U.S. cities have copied it (Waxler 1994).

Similar to the Grameen Bank, Women's World Banking (WWB) provides poor entrepreneurs access to finance information and markets. Founded in 1979, WWB has grown to provide over $150 million in loans to 500 000 women in 40 countries with a repayment rate of 98 per cent (Howells 1993).

The success of these two programs is believed to stem from the bottom-up approach, which provides the means for self-sufficiency. The result of micro-loan programs are as impressive as they are consistent. Businesses have expanded their incomes, increased job opportunities, provided gains in self-confidence, increased community participation and improved family functioning (Waxler 1994). World Bank programs which typically provide financial assistance do not result in self-sufficient businesses but instead instill a dependency on international assistance. It is for this reason that many of their programs have attracted criticism. One example of a World Bank program that has received much criticism is the completion of highway construction in the Brazilian Amazon, which was integrated with a World Bank settlement program (POLONORESTE). This settlement program resulted in large-scale destruction of the rain forest and much criticism (see Appendix A).

Whether micro-loans, education programs or subsidies are used as the means to support sustainable agriculture, it is the bottom-up approach, which includes farmers in the decision-making process, that can add to the success of the program. As these micro-loan programs have shown, self-sufficiency can motivate farmers to maintain sustainable practices. Although micro-loans have been a successful means of providing capital to smaller, poorer farmers, they are not currently necessary in order to support the adoption of sustainable agriculture in Ouro Preto do Oeste.

These loans are not necessary in the settlement of Ouro Preto do Oeste since the economic constraints have not been an issue in the adoption process. IPHAE (a local NGO) and labor organizations like APA have provided seedlings free of charge to farmers who have chosen to plant intercropped plots and/or join the organizations. Economic and capital constraints are commonly deterrents to the adoption of new agricultural technologies in developing countries (Miller and Tolley 1989, Holden 1993, Kebede *et al.* 1990, Jaeger and Matlon 1990, Norris and Batie 1987, Savagado *et al.* 1994, Rahm and Huffman 1984, Shakya and Flinn 1985). Since an initial cash investment is not necessary for the farmers in Ouro Preto do Oeste,[4] theory would predict that the adoption rate of the new technology would be high since there are labor and income benefits that are expected when sustainable agriculture is adopted. The adoption of sustainable agriculture is, however, very low in Ouro Preto do Oeste. Therefore other capital constraints such as stored wealth and labor are investigated in the regression analysis in Chapter 6.

3.5 THE ROLE OF ECONOMICS IN THE ADOPTION OF SUSTAINABLE AGRICULTURE

A bottom-up approach to the deforestation problem in the Brazilian Amazon is especially important because it addresses the problem at the local level. The typical farmer living in the southern borders of the Brazilian Amazon migrated to the area in response to government incentives and promises of a better life. The farmers are responsible for a major part of the deforestation in the area (Myers 1994), but destruction only continues because the farmers are burning the forest in order to provide food for themselves and their families. Anderson and Ioris (1992) found that small-scale farmers and rubber tappers on Combu

Island in the Brazilian Amazon had high and apparently sustainable economic return from the extraction of forest resources, such as fruits. Key to the success of these extractivists was the combination of proximity to a major market and appropriate resource management. Economic considerations are therefore an important part of the adoption process. Sustainable agriculture must be profitable if it is to be adopted by small-scale farmers who depend on the economic returns of agricultural products for their livelihoods.

ENDNOTES

1. Four of the most frequently used examples of intercropping on a one-hectare lot by these Rondônian farmers are 1) 300 pupunha, 300 cupuaçu, 200 banana, 50 mahogany, 100 Brazil nut and 200 other valuable wood trees; 2) 200 coconut, 200 cupuaçu, 200 banana and 100 valuable wood trees; 3) 150 graviola, 300 aracaboi, 50 Brazil nut, 50 mahogany, 100 acerola, 200 banana and 50 other valuable wood trees; and 4) 1000 coffee, 400 pupunha, 50 mahogany, 200 banana and 50 other valuable wood trees.
2. Technosolo, Porto Velho, Rondônia – a Brazilian and Dutch company – is mapping the soil types of the entire state and studying the effect of sustainable agriculture on soil integrity. The project began in 1997.
3. World Neighbors is a grassroots support program, based in the United States, which supports various programs in Central and South America to educate farmers about sustainable agriculture (Ruddell 1995, Adriance 1995).
4. The farming equipment necessary for the intercropping plots does not vary greatly from the equipment used in slash-and-burn agriculture. Farming equipment generally consists of low-cost hand tools. In addition, the use of fertilizers and herbicides do not play a role in intercropping.

Chapter 4

Addressing Market Failure Which Has Led to Tropical Deforestation With Discrete Choice Models

_____

4.1 THE ROLE OF ECONOMICS IN DEFORESTATION

Deforestation in the Amazon Basin of Brazil is the result of various market and government failures. These market and government failures account for the human and social pressures that occurred during the settlement of this area and the social and economic problems that exist in the country today. These circumstances that have led to deforestation in Brazil are the result of poorly defined property rights for land ownership, ill-conceived government policies that unfairly secure advantages to some farmers and ranchers through subsidies, and the divergence of social and private costs and benefits resulting from assigning different values to the forest. Although property rights are now secure in the area, and the governmental policies that supported deforestation are for the most part abolished, the deforestation trends are continuing in the Amazon region of Brazil.

   Market failure has occurred on global, national and local levels and has resulted in deforestation rates that are well above optimal levels. Independent of the level of occurrence, all of these market failures impact the decisions of local farmers and their land-use choices. A discussion of market failure and how it will be addressed follows.

25

4.2 MARKET FAILURES

Traditional economic solutions to market failures are the allocation of property rights, taxing the source of the market failure, subsidizing correct behavior, the use of tradable permits and the creation of markets by the government that induce proper market incentives. These solutions cannot be expected to yield an optimal solution (one where the marginal costs and benefits are equated) to Amazon deforestation because of the complexity of the situation. Each of these solutions, individually, may only address a portion of the problem. For example, when property rights became more defined, the rate of deforestation in the Amazon did not decrease. Also, in the state of Rondônia, as they gain titles, farmers are continuing to burn their lots with no apparent concern for the future. Taxing and/or the use of marketable permits is not feasible in this area of the world since the economy is not well developed and the farmers are very poor.

Subsidizing correct behavior, such as the use of sustainable agriculture, is a potential solution. However, as is pointed out in Chapter 3, financial assistance of small enterprises and farmers in developing nations has often been unsuccessful (Waxler 1994). Instead, programs that support self-sufficiency have been highly successful (Fragasso 1995, Southgate 1994, Waxler 1994, Howells 1993). Therefore, successfully subsidizing sustainable agriculture would require the use of educational programs to teach farmers how to plant and maintain these systems and would consequently also require a substantial commitment. Without these education programs sustainable agriculture will most likely fail. IPHAE[1] representatives studied sustainable agricultural systems in Ouro Preto do Oeste. These systems were subsidized by the government[2] and put in place without educational support. The Technosolo study found that these systems were not maintained and were unsuccessful in producing any goods that could be harvested and sold. The market failures that have resulted in tropical deforestation can be classified on global, national and local levels. These classifications are discussed in the following sections.

4.2.1 Global Market Failure

Global market failures have occurred as global and international economies have placed benefit values on the standing forest. These global benefits such as nonuse and existence values have not been reflected in global markets. As mentioned in Chapter 2, one example of a cost that deforestation has imposed on the global level is the increase in carbon emissions, which add to greenhouse gases. Pearce and Brown (1994) estimate that the carbon damage caused by the conversion of primary forest for agricultural purposes is about $4000–$4400 per hectare of tropical forest.[3] At the same time the private value of land was found to have an upper bound of $440 per hectare in Ouro Preto do Oeste in this research endeavor. If the existence and nonuse values that global entities place on the standing forest are included in the value of each hectare of forest, the total would be well over $4400 per hectare. The disparate values between global and private land prices define the global market failure caused by tropical deforestation.

4.2.2 National Market Failure

The national-level market failures that have led to deforestation in Brazil are related to a nationally increasing trend in poverty and overpopulation in the southern and northeastern regions of the country and the national debt (Kahn and McDonald 1995, Frohn *et al.* 1990, Moran 1990). Various failed governmental programs aimed at ameliorating these problems have led to deforestation in the tropical region of the country. These government failures began in the 1960s with the launching of the colonization program known as 'Operation Amazonia'. This settlement program began the development of the Amazon region and ended with many unsuccessful farm establishments and the beginning of slash-and-burn agriculture in the Amazon region by small-scale farmers.[4]

4.2.3 Local Market Failure

The market structure of the Brazilian farm settlements of the Amazon are based on annual crops that can be bought and sold easily in well-developed markets. Unfortunately, it is the growth of the market for these products that destroys the rainforest. Products native to the area such as fruit, nuts and honey do not have well-developed markets and are, therefore, much more difficult to sell. Trading native products could provide incentives to maintain the rainforest instead of cutting it down.

In a study of non-timber forest products (fibers, canes, resins, oils, fruits and nuts) in Ecuador, Grimes *et al.* (1994) find that the present value of the net revenue from the non-timber forest goods is $2830. The value of non-timber forest goods was found to be significantly higher than other forest uses in the area, including agriculture (the present value is estimated to be less than $500 per hectare). The value or price of these goods is subject, however, to supply and demand. Organized markets are necessary in order to sell these goods.

One reason that the rate of return of the forest has been undervalued on the local level is because of missing markets (Pearce and Brown 1994). There are many products and functions of the forest that are not being valued by the farmers who cut and burn the tropical forest. Products such as nuts, fruits and medicinal plants are not being harvested because there are no organized local, national or state-wide markets to sell these products. Even if these markets did exist, the lack of a developed transportation system prevents the bringing of these products to market. Scientists and non-governmental organizations are working to find products other than rubber and Brazil nuts that can be harvested and sold in new and developing markets. Some of the scientists and organizations are addressing the overwhelming problems of communication and transportation (Holloway 1993). When these products and functions are not included in the calculation of the value of the forest, the opportunity cost of cutting it down is greatly reduced.

In total, these failures prevent Brazilian farmers who live in the Amazon from making the best long-term choices for Brazil and the rest of the world (Pearce and Brown 1994, Kahn and McDonald 1995, Miller *et al.* 1993). The economic theory behind the forest is simple: if it can be shown that forests are of more value standing than felled, then they are more likely to be preserved (Corry 1993). The inequality between private and social values makes the solution more difficult than the theory suggests. The farmer who cuts down the forest places a private value of standing forest at a much lower level than the social value that reflects the values of biodiversity, forest products, and native plants and animals. The felled forest is often, unfortunately, of more value to the private farmer.

4.3 ADDRESSING MARKET FAILURE

Market failures occurring on global, national and local levels make it difficult to find an optimal solution to deforestation on all levels simultaneously. The optimal rate of deforestation is determined as the level of deforestation where the social and private benefits of the standing forest are equal. On each level, market forces are not securing the economically correct balance of land conversion and land conservation (Pearce and Brown 1994). This situation occurs because, as with all public goods, those who convert or destroy the forest do not have to compensate those who value the standing forest.

The economic solutions mentioned earlier do not result in zero deforestation, but rather in economic balance of conversion and conservation, resulting in the optimal rate of deforestation. Simultaneously solving for an optimal rate of deforestation on all of these levels is not possible since benefits differ across tiers. Kahn (1995) points out that deforestation rates that are optimal on a national level may not be optimal from a global point of view because the country will only consider the domestic costs and benefits and not take into consideration global nonuse or existence benefits. International-level benefits of the tropical forest include nonuse and existence values. National benefits may include timber for trade and the conversion of forest to coffee plantations to support national trade balances and the national economy. On the local level, benefits of the forest include conversion to farms and ranches to support subsistence living. These differing optimal levels of deforestation can be represented in Figure 4.1. In this figure the marginal costs are presented for the private (MPC), national (MSC$_N$) and global (MSC$_G$) levels where MPC stands for marginal private cost and MSC stands for marginal social cost. The intersection of the different marginal costs and the marginal benefit (MB) indicate the optimal private (D$_P$), national (D$_N$) and global (D$_G$) levels of deforestation. The optimal level of deforestation decreases as more uses of the forest are included in the marginal cost measurements.

Since an optimal solution cannot be found to help reduce deforestation, a second-best solution that induces a change in behavior can result in an increase in welfare both on the local level and in society. The rate of

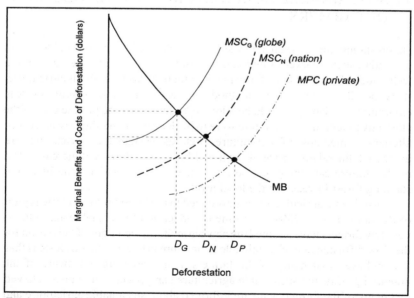

Figure 4.1 Global, National and Private Optimal Levels of Deforestation

deforestation can be decreased in some regions of Brazil by altering traditional farming techniques and using sustainable agricultural practices. These methods could provide the financial support of farmers for an extended period of time while increasing the amount of time spent cultivating on each lot (Dale *et al.* 1994, Goodland 1980). Sustainable agricultural practices have been used successfully by the indigenous people of the area for centuries (Kahn 1995, Anderson 1993). The practice is therefore supported by many professionals as a sustainable technique and a promising alternative to slash-and-burn agriculture, although it is not a cure-all for the deforestation problem.

Increasing pressures from population, in an increasingly more densely populated area, may contribute to the success of sustainable agriculture. Most small farmer deforesters of the 1990s have, however, come from within the Amazon region. The greatest threat to deforestation in the Amazon today is, therefore, from intra-regional migration. According to Almeida and Campari (1995) the issue is no longer how to prevent farmers from migrating to the Amazon, but how to ensure that the farmers who occupy Amazonian land remain on their farm lots, reducing migration into the interior of the Amazon where deforestation has been minimal.

4.4 SUSTAINABLE AGRICULTURE AND THE DECISION OF FARMERS

Deforestation may be reduced by solving market failure at the local level. In particular, local market failure can be resolved by altering the agricultural techniques of a majority of farmers who practice slash-and-burn agriculture. It is possible that a new method of agriculture that results in less environmental damage would be adopted by farmers if the private cost of the alternative method of agriculture were to be decreased. As the private cost of alternative methods of agriculture such as sustainable agriculture are decreased, the relative price of slash-and-burn agriculture will increase. The market failure can, therefore, be internalized as the relative benefits of the standing forest increase on the local level.

Sustainable agriculture was introduced to the Ouro Preto do Oeste region of Rondônia in the 1980s. Since the introduction of sustainable agriculture, very few farmers have adopted the new agricultural technique. One reason for the slow diffusion rate of sustainable agriculture in Ouro Preto do Oeste is that it is believed that many of the farmers in this area are not aware of the potential profits that sustainable agriculture can provide. It is also believed that many of the farmers who are informed about sustainable agriculture and the potential profits may not be willing to take the risk of switching to an unfamiliar method of farming. For both of these reasons, the lack of knowledge and risk-averse behavior, many farmers have not adopted the new method of farming. Relatively high discount rates (Ehui *et al.* 1990) increase the incentive to use slash-and-burn agriculture.

Under the present conditions, farmers are making rational decisions. Slash-and-burn methods may be more profitable in the short run, but as the soil quality declines, profits fall (Goodland 1980). Short-term decisions and the dependence on farming for their livelihood have made sustainable methods of agriculture often unaffordable. Since most of these farmers do not have access to capital or loans, the option to burn and plant is often the choice of farmers when outside assistance with more sustainable methods of farming is not available.

There is a risk involved in adopting sustainable agriculture since there are no fully developed markets to sell the products that are grown – such as cupuaçú, pupunha, açaí and acerola, to name a few – while there are well-developed markets for annual crops. The establishment of infrastructure and markets is key to maintaining a healthy economy that can support the commercialization of forest products. In addition there is a risk involved with income, since it is not known whether these products can be easily sold or not.

Another deterrent to adopting sustainable agriculture for these farmers is that a large initial capital investment is required if outside assistance is not provided. The seedlings necessary to plant a one-hectare sustainable agricultural plot are costly and can range from $450 to $900. This expense can represent up to 33 per cent of the annual income of the farmer in Ouro Preto do Oeste. In addition, changing to a new method of farming will require investments in terms of education and finances. A great deal of information and training is necessary in order to switch to new agricultural methods. It is therefore important that policies address these issues and help to reduce the risks that are involved for the farmers.

4.5 DIFFUSION AND ADOPTION LITERATURE

Studies that have addressed local market failure in developing countries by the adoption of new techniques of agriculture have investigated a wide variety of technologies. A common theme throughout these papers is that the adoption of new techniques that increase welfare and/or income often occur at low diffusion rates (Holden 1993, Kebede *et al.* 1990, Sands 1986, Opare 1977). Many farmers are either not willing to adopt the new technology or are constrained by capital resources and cannot adopt the potentially superior technology. This observation was also made in Ouro Preto do Oeste. Only 12 per cent of the farmers, from a random sample, used sustainable agriculture even though the technique was shown to improve the welfare and income of those farmers who used it.

The adoption of technical innovations in agriculture has attracted considerable attention among developmental and environmental economists since the majority of the population of developing countries derives its livelihood from agricultural production. New technologies in agriculture can provide the opportunity to increase production, yield and income in these developing countries (Feder *et al.* 1985). The low rates of adoption in developing countries are contrary to what economic theory might predict, since this asserts that the technology yielding the highest benefits will be adopted. Conventional diffusion analysis asserts that a rational individual will adopt a technology that is more profitable or will provide a greater utility. These technologies are often not being adopted because of a lack of information about the technology, a shortage of capital assets and of the means necessary to transport inputs, and because of a local custom or tradition that may be associated with the older technology or method of farming.

Models that estimate the adoption of new techniques and technologies in developing countries have focused on different issues from those for developed countries. The adoption-decision literature was first introduced in terms of firms' decisions in developed countries. Much of the literature, which has focused on the adoption decisions of firms in the U.S., disregards social factors and concentrates on the issues of market size, market characteristics, industry size, risk and the characteristics of the new technology (Hannan and McDowell 1984, Oster 1982, Griliches 1957). Unlike developed countries where well-formed industries exist and information is relatively easy to obtain, the adoption of new technologies and farming methods in developing countries is often influenced by various social factors in addition to economic ones. It is therefore essential that social factors are considered in the modeling process. For example, Kebede *et al.* (1990) found that in Ethiopia the adoption of new technologies in farming is determined both by economic and social variables and that social variables are crucial to the probability model. Adesina and Zinnah (1993) show that the perceptions of technology and the social impacts that the new technology would make are major factors in determining adoption for farmers in Sierra Leone. And, as Holden (1993) has found, farmers in Zambia had been given the choice to switch from the more traditional slash-and-burn agriculture to a new maize processing system that would reduce the labor requirement by as much as 40 per cent and generate higher profits. The adoption of the superior technology was not undertaken by many farmers because the traditional means are supported by the social system. In addition, Holden found that in areas where there was a dependence on farm income, and alternative off-farm income was not available, the risk involved in adopting the new technology increased.

The farmers in the Ouro Preto do Oeste region of Rondônia have various choices about how they can farm their land. They can cultivate the more traditional European crops such as corn, rice, beans, manioc and coffee; grow perennial crops native to the area, such as pupunha, cupuaçú and valuable woods; convert the land to pasture and raise cattle; or choose any combination of these land uses. The choices that these farmers make depend partially upon their past farming experience, assets and information. These considerations can be represented using a probability model explaining the choices that a farmer can make between using sustainable agriculture and using slash-and-burn agriculture.

The development and analysis of this model is discussed in detail in Chapter 6. The analysis tests the role that information, education and local organizations play on farmer choice. It is suggested from the literature review that the more highly educated farmers who have knowledge of sustainable agriculture and participate in local farm organizations are more likely to adopt sustainable agriculture. Variables such as education level, knowledge of sustainable agriculture, union participation and others are investigated in the analysis. This chapter discusses models that estimate the probability of adopting a new technique of agriculture and provides the motivation for the model used in the analysis.

4.6 THE INTRODUCTION OF DIFFUSION AND ADOPTION TO ECONOMIC LITERATURE

The diffusion of innovations in agriculture was introduced to the economic literature with the seminal work of Griliches (1957). The literature now contains a broad spectrum of approaches and models to the diffusion process and the adoption decision of the farmer. Essentially these approaches differ from one another in their degree of, concern with, and emphasis on the relative importance of different factors and variables likely to influence the adoption of an innovation (Argarwal 1983).

The adoption model developed by Griliches was further refined by Mansfield (1971). Mansfield defined the pattern of adoption by a logistic or S-shaped adoption curve of increases in the number of farmers who adopt the new technology, plotted against time (Figure 4.2). When an innovation is first introduced, a firm may be uncertain regarding its profitability and success should the firm adopt it (point A). This uncertainty is reduced over time as the

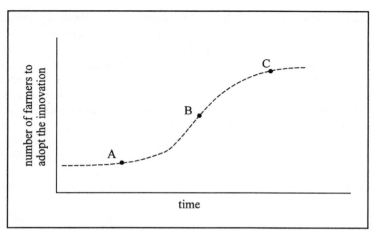

Figure 4.2 Logistic Diffusion Curve

information about the innovation, and the number of firms that adopt it, accumulate (point B). As the information increases, the probability that the innovation will be adopted increases, resulting in an S-shape diffusion curve or a logistic (learning) curve.

The rationale for this diffusion pattern through time is that the adoption of an innovation accelerates initially as it becomes more widely known and competitive pressure compels its adoption. Eventually the rate of adoption slows as the number of firms not employing the innovation declines (point C). When the technology is first introduced, the prospect of an innovation can offset the risks associated with the use of a new process, and, as more firms adopt it, the risks are further reduced. Uncertainty is reduced in proportion to the number of firms already using the technology. The speed of the response is assumed to be related to the size of the stimulus.

Mansfield tested the model on a sample of 12 process innovations in four industries: bituminous coal, iron and steel, brewing and railroads. He found that his model could support the logistic diffusion pattern of adoption very well. He also explored inter-industry diffusion and found that the diffusion patterns differ among industries, finding a more rapid rate in more competitive industries. The next section contains a discussion of studies that investigate agricultural technologies in developing countries and that are based upon this diffusion theory.

4.7 OVERVIEW OF AGRICULTURAL ADOPTION LITERATURE

The seminal work of Griliches (1957) first introduced the adoption choice by farmers to this literature. In this paper Griliches studied the factors responsible for the wide cross-sectional differences in past and current rates of the use of hybrid corn in the United States. He derived a diffusion path of adoption and concluded that 'the process of innovation, the process of adopting and distributing a particular invention to different markets and the rate at which it is accepted by entrepreneurs are amenable to economic analysis' (p. 501). Since the Griliches paper, studies have distinguished between two and three categories into which the adoption choice studies fall (Adesina and Zinnah 1993, Nowak 1987, Argarwal 1983), but since then four distinct paradigms have surfaced that can categorize the adoption decisions of farmers in developing countries. These categories include the human capital framework, the innovation diffusion framework, the economic constraint framework and the adopter perception framework (see Table 4.1).

4.7.1 The Adoption Choice Frameworks

The human capital framework and innovation diffusion framework were the first to be explored in the literature. Griliches' paper (1957), the first study of the adoption and diffusion of new technologies, falls into the human capital category. The economic constraint framework was soon to follow (Adesina and Zinnah 1993, Nowak 1987, Argarwal 1983). The adopter perception framework is the most recent approach to be examined (Adesina and Zinnah 1993). A review of these four adoption choice frameworks is provided in the following four sections. Each of the frameworks is first defined, followed by a review of some studies that fall in the appropriate category. The model of adoption of sustainable agriculture by small-scale farmers in Ouro Preto do Oeste investigates the role that each of these frameworks can make in the adoption decision. The importance of issues such as human capital constraints, economic constraints, the role of information and farmer perceptions are investigated in the model.

Although the distinction between paradigms is not always obvious, the category in which each study falls is based on the factor that influences the adoption decision the most. In addition, a few examples of how studies of this type may be extended past the adoption decision are examined in section 4.7.6. As well as providing a description of studies, Table 4.1 arranges the studies discussed in the following sections by category.

4.7.2 Human Capital Framework

The human capital framework suggests that the adoption of a new technology by an individual (firm or person) is determined by characteristics unique to that individual (Strauss *et al.* 1991, Lin 1991, Akinola and Young 1985). In the case of farmers these include education, age, experience and family size. According to this framework only those individuals who fit a certain character type will adopt the new technology (i.e. those in a certain age and education cohort).

The human capital framework was inspired by Schultz (1964), who argued that the frequent introduction of new technologies results in a disequilibrium and suboptimal use of inputs and technologies, even though in static agriculture resource allocation is efficient. Thus, changes in the technological environment increase the farmer's ability to perceive, interpret and respond to new events in the context of risk.

Table 4.1 The Adoption and Diffusion of New Farm Technologies – Studies in Developing Countries

	human capital model	innovation diffusion paradigm	economic constraint model	adopter perception model
Assumption	the adoption of a new technology is determined by characteristics that are unique to that individual	the access to information is a key factor to determining adoption decisions	economic constraints are the major determinants of adoption behavior	the perceived attributes of innovations condition adoption behavior
Studies	Strauss *et al.* (1991), Lin (1991), Akinola and Young (1985)	Hussain *et al.* (1994), D'Souza *et al.* (1993), Feder and Slade (1984), Opare (1977)	Miller and Tolley (1989), Holden (1993), Kebede *et al.* (1990), Jaeger and Matlon (1990), Norris and Batie (1987), Savagado *et al.* (1994), Rahm and Huffman (1984), Shakya and Flinn (1985)	Saha *et al.* (1994), Adesina and Zinnah (1993), Gould *et al.* (1989), Lynne *et al.* (1988)

Other studies that fall into the human capital framework are Strauss *et al.* (1991), Lin (1991), and Akinola and Young (1985). Strauss *et al.* (1991) studied the role of education and extension programs in the adoption of technology of rice and soybean farmers in the central-west region of Brazil. The authors found that farmer education has a positive impact on the diffusion process of both upland rice and soybean technology adoption. Lin (1991) used the diffusion of hybrid rice in the Hunan province of China in order to examine the effects of education on the adoption of the new technology. The results of the Lin study are consistent with the hypothesis that those farmers with higher education are more likely to adopt hybrid rice. In addition, Lin found farm size to be positively and significantly related to adoption, suggesting that smaller farms in China may show resistance to technological change. Akinola and Young (1985) studied the decision process of individual cocoa farmers in Nigeria in order to determine the factors that influence their decisions, which are made in terms of land use and cocoa production. The authors found evidence that income, size of the family, age and distance from the market have the greatest effect on the adoption of new cocoa techniques. In particular, they found that wealthier younger farmers were more likely to adopt the new technology.

The human capital framework suggests the need for consideration of farmer and household characteristics. As the studies that have been reviewed have shown, characteristics such as education, age and family size can play important roles in the adoption of new technologies by small farmers in less developed countries. These characteristics are, therefore, included in the analysis demonstrated in Chapter 6. In the analysis education level and age play significant roles as in the studies by Schultz (1964), Strauss *et al.* (1991), Lin (1991), and Akinola and Young (1985). The human capital paradigm asserts that younger farmers are more likely to adopt new techniques and technologies. The analysis in Chapter 6 may or may not support this conclusion of the human capital paradigm; therefore, factors other than farmer specific characteristics are also investigated to determine the rate of adoption of sustainable agriculture in Ouro Preto do Oeste.

4.7.3 Innovation Diffusion Paradigm

The innovation diffusion framework asserts that access to information is a key factor in determining adoption decisions. The problem of adoption is reduced to communicating information on the technology to the potential end users. According to the framework, programs that emphasize the use and means of communication and on-farm prototypes can influence risk-averse non-adopters to adopt (Hussain *et al.* 1994, D'Souza *et al.* 1993, Feder and Slade 1984, Opare 1977). Therefore, according to this framework, a rational farmer will adopt a superior technology when information is obtained about how it is used, and obtained, and/or information is gained as to its expected results. This information may be passed on through means such as government or non-government-supported programs, independent research or advice from neighboring farmers.

The innovation diffusion paradigm concentrates on the impact that knowledge of the technology can make on the adoption decision. Studies that fall under this framework are Hussain *et al.* (1994), Feder and Slade (1984), and Opare (1977).

Hussain *et al.* (1994) used the diffusion framework to determine the impacts of training and visit extension programs on the adoption of new wheat technologies by farmers in the Pakistan Punjab. The authors found that when the quantity of extension programs increased, quality did not necessarily increase equally. Even so, an increase in the number of extension programs increased farmers' knowledge and adoption of the new technology.

Feder and Slade (1984) determined how information affects the adoption decision of farmers in India and found that as farm size increases so does the information acquired, at a cost, by the individual farmers. As information increases so does the adoption of the new technology. Therefore farmers with larger farms, with better access to information or with more human capital will adopt the new technology earlier than other farmers.

Opare (1977) investigated the adoption of new cocoa-growing techniques in Ghana and found that the most important predictor of adoption is growers' knowledge of the principles underlying the recommended practices. The farmers who best understand the new technology are the most likely to adopt it.

All three studies reviewed here support the theory that information and knowledge of a new technology can increase the rate of adoption. As the knowledge of the new technology increases, the rate of adoption is expected to increase. The studies also point out that the spread of information in rural areas of developing countries can be difficult. Information about sustainable agriculture is a relevant issue in Ouro Preto do Oeste. It was found that only 24 per cent of the farmers who were interviewed had heard about or had knowledge of sustainable agriculture. Of those who knew about sustainable agriculture, 89 per cent chose to adopt the new technology. These percentages suggest that information plays a vital role in the adoption of sustainable agriculture in Ouro Preto do Oeste. The role that information plays in the adoption of sustainable agriculture is therefore included in the regression analysis in Chapter 6.

4.7.4 Economic Constraint Framework

The economic constraint framework contends that economic constraints reflected in asymmetrical distribution patterns of capital and labor endowments are the major determinants of adoption behavior (Miller and Tolley 1989, Holden 1993, Kebede *et al.* 1990, Jaeger and Matlon 1990, Norris and Batie 1987, Savagado *et al.* 1994, Rahm and Huffman 1984, Shakya and Flinn 1985). Some constraints that may fall into this category for farmers are farm size, wealth, debt, capital equipment and access to financial capital. According to this paradigm, when capital and labor constraints are not binding, adoption will occur.

Studies that fall under this framework include Miller and Tolley (1989), Holden (1993), Kebede *et al.* (1990), Jaeger and Matlon (1990), Savagado *et al.* (1994), and Shakya and Flinn (1985). All six studies found that access to capital in the different forms (fertilizer, workers, money capital, farm equipment and stored wealth) encourage the adoption of new technologies and crops.

In all these studies the access to capital, in one form or another, increases the rate of adoption. For example, Holden (1993) studied the adoption of new maize farming and fertilizing techniques in northern Zambia. Holden found, from the examination of six villages, that those areas which experienced rapid expansion of new maize and fertilizer technology had a favorable price ratio between fertilizer and maize output, limited access to off-farm employment, and a supply of capital equipment to handle fertilizer supply and maize marketing. These results imply that those farmers who cannot find off-farm employment are full-time farmers, and those who have access to capital equipment are more likely to adopt the new technology. The economic constraint framework introduces the restrictions that the access to capital goods or equipment can pose, especially for small, poor farmers in developing countries.

The analysis performed in Chapter 6 investigates the role that capital constraints play in the adoption of sustainable agriculture in Ouro Preto do Oeste. Factors such as debt, capital equipment and per capita income are investigated in addition to capital constraints such as labor availability and wealth.

4.7.5 Adopter Perception Framework

The adopter perception framework suggests that the perceived attributes of innovations condition adoption behavior (Saha *et al.* 1994, Adesina and Zinnah 1993, Gould *et al.* 1989, Lynne *et al.* 1988). The farmers' preconceived judgments of technology and their attitudes toward it determine their adoption choice. These perceptions can be influenced by a variety of social characteristics such as traditions and customs. This framework suggests that adoption is more likely to occur when local traditions and customs are accounted for.

Papers that fall in this category include Saha *et al.* (1994), Adesina and Zinnah (1993), and Gould *et al.* (1989). One study that focuses on incomplete information in the adoption process is Saha *et al.* (1994). The authors developed a framework of technology adoption under incomplete information dissemination and output uncertainty in order to determine the role that expected income gains from the new technology (among other variables) play in Texas dairy farmers' adoption of a new technique capable of increasing cows' milk production by 10–20 per cent. They focused on the analytical and empirical implications of incomplete information in the adoption process. They argued that producers' choices are significantly affected by their exposure to information about the new technology and determined that the farmers' expectations of yield and adoption costs are the sole determinants of the adoption decision.

Adesina and Zinnah (1993) found similar results when they tested the influence of farmers' perceptions of technology-specific characteristics on the technology adoption decision of mangrove swamp rice farmers in Sierra Leone, Africa. Farmers' perceptions of the technology-specific attributes of the modern rice varieties were determined to be the major influences in determining adoption.

Perceptions of new technologies and techniques can play a major role in developing countries where customs and traditions dominate economic and social behavior. New techniques and technologies that conflict with local traditions are not likely to be adopted by many farmers. The analysis in Chapter 6 does not investigate this paradigm in terms of attitude toward sustainable agriculture. Attitude or feelings toward a particular technique of agriculture are difficult to measure. Instead, the role of farming experience is investigated. It is expected that the farmers who have used slash-and-burn agriculture for many years are more likely to have adverse feelings toward different methods of farming.

4.7.6 Extensions of the Adoption Literature

The following articles are a small sample of how the adoption choice literature may be extended past the adoption decision. One possible extension is that simulations can be made based upon the probability results and regional characteristics and statistics to determine the impacts of the new technologies on local economies, welfare distribution and the environment.

Renkow (1993) used a multi-market model of technical change in food production in order to simulate the long-run economic impacts of various technological programs available for wheat farmers in Pakistan. Renkow was interested in determining whether policies that promote new wheat technologies are evenly distributed among income groups. Using data collected in a 1987 survey in the Punjab province, Renkow determined that the traditional programs that promote adoption of the new technologies are reasonable. Capital assets of the family prior to adoption do affect the benefits received. Net consuming households, those which purchase more farming products than they produce, stand to be the major beneficiaries of productivity increases resulting from the technological progress.

Coxhead and Warr (1991) investigated the distributional effects of technological changes using a computable general equilibrium model to simulate three hypothetical forms of technological change: factor-neutral–labor-saving, capital-using and labor-using land saving. The nature and magnitudes of these technological changes are illustrative of individual technical innovations in developing country agriculture. The model distinguishes between commodities traded internationally and those that are not, technologies that are fixed from mobile factors of production and technologies that contain technical changes from different factor biases. The simulations suggest three conclusions: farmers who do not adopt the new technology may be forced to do so later on due to losses, other sectors may suffer when technological improvements are made in the agricultural sector, and gains from production may be constrained when there is a lack of the necessary complementary inputs.

4.8 THE ROLE OF ADOPTION CHOICE MODELS IN DEVELOPING COUNTRY STUDIES

The four major frameworks in the literature on technology adoption each emphasize different factors that are important in the adoption choice. This research will help determine which factors are the most important in the adoption of sustainable agriculture in the Ouro Preto do Oeste region of

Brazil. A key to understanding why more farmers do not choose new technologies in developing countries may be a lack of capital assets and requirements, infrastructure problems, governmental intervention and a wide variety of other considerations that are unique to each region, as the four paradigms discussed suggest. In addition, the analysis in Chapter 6 investigates the role that farm unions and associations play in the adoption of sustainable agriculture. These organizations have served as a means of spreading information on sustainable agriculture. Unions provide educational programs on how to plant the intercropping systems and harvest honey, and have provided, free of charge, the seedlings used in intercropping.

The analysis of farmers in Ouro Preto do Oeste is interesting since no initial capital investment is required for those farmers who receive free seedlings from local organizations and unions. Economic theory predicts that capital investment and risk-averse behavior are the major deterrents to the adoption of new technologies. According to this theory, the adoption of sustainable agriculture should, therefore, be relatively high in Ouro Preto do Oeste since the capital investment is zero and the risk is relatively low. The risk is reduced as farmers learn about the experiences of other farmers who have adopted sustainable agriculture at union-organized workshops. The diffusion of sustainable agriculture is, however, low. One major reason for the low diffusion rate could be that the area is lacking a developed communication system. Technical feasibility and increases in yields, income and/or production are not enough to guarantee adoption in developing countries. Information and the diffusion of information also play major roles and for this reason these variables are investigated to determine the part that they play in the adoption of sustainable agriculture in Ouro Preto do Oeste.

This chapter has investigated the market failures that have led to deforestation in the Brazilian Amazon. The analysis determines that addressing the market failure on the local level can reduce deforestation rates rapidly if sustainable agriculture is adopted by many of the farmers who use slash-and-burn agriculture. This chapter continues by reviewing how the economic literature on the adoption of new farm technologies in developing countries has emerged from the seminal work of Griliches (1957) to include not only human and economic characteristics, but also economic constraints, information constraints and farmer perceptions. In addition to progress in the economic and social considerations, the modeling techniques that are used to illustrate the dichotomous choice of adopting a new technology or not have also advanced from simple ordinary least squares (OLS) models to probit and logit probability models, and to more complex discrete choice models such as tobit and Heckman models. The advancement of the modeling techniques in this literature is discussed in Appendix D.

ENDNOTES

1. IPHAE is a non-governmental organization (NGO) with headquarters in Porto Velho, Rondônia. This NGO assisted farmers in planting intercropping plots in three settlements of Rondônia: Ouro Preto do Oeste, Ariquemes and Jamari.
2. Based on an unpublished study that analyzed intercropping programs initiated in 1990 by IPHAE (an NGO with headquarters in Porto Velho). Pieter Van Der Veld, an independent Dutch agronomist, was employed by IPHAE to administer the study. (Pieter Van Der Veld and Monique Berensden, personal communication, November 1996).
3. This estimate is based on a study by Fankhauser (1993), which suggests a central value of $20 of damage for every ton of carbon emitted.
4. See Appendix A for a detailed account of Operation Amazonia.

Chapter 5

Data Collection and Analysis

5.1 DATA USED IN THE ANALYSIS

The data used to estimate the probability of adopting sustainable agriculture and the extent of adoption were collected in Ouro Preto do Oeste, Rondônia, from small farms. The goal of the data collection was to develop information that allows for the testing of hypotheses about the determinants of participation in sustainable agriculture. The data set is used to determine which factors can best promote the adoption of sustainable agriculture by more farmers. The data are a micro-level grouping of small farm information collected using an oral survey conducted at the homes of a majority of the farmers. A select number of the variables collected are used in the regression analysis, which is described in Chapter 6. This chapter describes the data collection process, survey design and data in detail.

5.2 SURVEY DESIGN

The data used in the analysis of farmer behavior in Ouro Preto do Oeste, Rondônia, were collected over a five-month period and are divided among two different groups. The first group was collected according to stratified random sampling, and the second was collected from farmers who participated in the local farmer union, the Association of Alternative Producers (APA). The survey, designed to be given orally, was written according to the guidelines in Fink and Kosecoff (1985) for cross-sectional data collection. An oral questionnaire was chosen since most of the people who were interviewed do not read or write. A majority of the questions are closed-ended with several choices, since these are more precise and, therefore, easily transferred into data that can be used in regressions analysis. All terms are defined fully in the questionnaire, which is attached as Appendix B in English and Appendix C in Portuguese.

The survey was preceded by an introduction by the assistant, Walmir de Jesus, a farmer from the area, who explained the purpose of the survey, described the research to be conducted with the information gathered and identified the interviewer and her qualifications. There was no compensation offered to respondents. Objective facts about the family and household were asked first and the more subjective questions were asked afterward. The survey ended with closed-ended questions in category scales (a lot of influence, some influence, no influence) that could be answered quickly after the end of a long interview (a range of 30 to 60 minutes).

Standardized interviewing techniques (Fowler and Mangione 1990) were used in order to minimize interviewer-related error. The questions were read exactly as worded. Participation by the assistant, Walmir de Jesus, a native to the area, was done in the same manner in each interview. The answers only reflected what the respondent said. Differences in interviews may have occurred between early and late interviews as the interviewer learned to speak Portuguese better. The assistant, fluent in Portuguese, had to intervene more often in the early interviews.

The surveys developed by Whitcover and Vosti (April, June 1996) for research in the Brazilian Amazon were used as guidelines for questions about native trees and local farm practices, such as the differences in harvest in the rainy and dry seasons and the names of native trees and fruits. In addition, local farming terminology was aided by a review of these surveys.

The Whitcover and Vosti surveys (April, May 1996) were conducted as part of a project by the International Food Policy Research Institute (IFPRI). The institute's research program reflects a worldwide collaboration to identify and analyze alternative national and international strategies for meeting the food needs of the developing world on a sustainable basis, with particular emphasis on low-income countries and the poorer groups in these countries. Those surveys were administered to approximately 150 small- to medium-scale households in two colonization projects of western Brazilian Amazon in the states of Acre and Rondônia. The purpose of the project was to provide socioeconomic information on land use patterns, the factors hypothesized to influence land use decisions and demographic characterizations of the households.

The final form of the questionnaire administered in this study evolved through approval by a University of Tennessee research and survey committee, review of the Whitcover and Vosti surveys (April, June 1996), and collaboration with University of Amazonas socioeconomic researchers and Brazilian surveyors in Ouro Preto do Oeste. The survey was pilot-tested on farmers who participate in the University of Amazonas agricultural programs and at the labor union of rural workers headquarters in Ouro Preto do Oeste (Sindicato dos Trabalhadores Rurais – STR). A total of 12 pre-tests were administered. These pre-tests were followed by discussions and modifications to the document to answer the realities met in the field.

5.3 DATA GATHERING

It was determined that about 150 farmers and their families would be interviewed in the city of Ouro Preto do Oeste. This total represents 4.49 per cent of the populated lots in the city. This total was chosen in order to collect a representative sample of the area (in terms of population of the six municipals of Ouro Preto do Oeste) considering time and budget constraints.

The interviews were conducted over a five-month period (September 1996–January 1997). The limited time period permitted the desired number of interviews to be collected, but did not allow for many interviews to be conducted at the home of the farmers in January, the height of the rainy season. This is when many of the Association of Alternative Producers (APA) member interviews were conducted at APA headquarters. Budget constraints allowed for the interviewer to be assisted by a local farmer, who was paid, and for a jeep, which was rented, but could not allow for the interviews to be extended through the assistance of additional interviewers. Additional interviews were collected in each of the six municipals beyond the goal number of interviews, to allow for some of the interviews to be dropped if the analysis determined any observations to be outliers. These additional interviews brought the total number of the city interviews to 171.

The first data set that was collected is a stratified random sample of 171 farmers who live in Ouro Preto do Oeste, which represents 1.47 per cent of the rural population, 3.07 per cent of total farm lots and approximately 5.12 per cent of the populated lots. The data set is stratified in order to represent each of the six municipals equally according to the number of lots residing in each. The advantages of this type of sampling over simple random sampling are that it is more precise and it permits the surveyor to choose a sample that represents the various groups and patterns in the desired proportions (Fink and Kosecoff 1985). The second data set is a sample of 25 farmers who participate in the labor union APA. The data represent 64.10 per cent of the active members. The population of the Ouro Preto do Oeste is 97 941 according to 1994 census updates, 11 617 of which is rural and 86 324 urban (Editoria Turística 1994). The city is divided into six different municipals: Ouro Preto, Valé do Pariso, Nova União, Mirante da Serra, Urupá and Teixeropolis (see Figure 5.1 for a map of these six regions). The municipal of Ouro Preto serves as the city center, and contains the post office, bank, and farm and labor unions. This municipal is developed to a greater degree than the remaining five municipals. Ouro Preto municipal has a well-developed infrastructure in the forms of organized markets and stores as well as electricity and telephone services. The five remaining municipals have well-developed markets in the center of the municipals, but are lacking in electricity and telephone services. The six different municipals are divided according to the year of settlement, soil type and geographical boundaries.

The 150 interviews were divided among these six different regions according to population density and number of farm lots in the region in a stratified random sample. The interviews were divided according to the number of lots in each region, in an attempt to represent each region spatially in terms of populated lots, since the purpose of this research is not only to represent the farmer but also the land use in the city of Ouro Preto do Oeste.

Figure 5.1 Map of Ouro Preto do Oeste: Division Among the Six Municipals

Sustainable Agriculture in Brazil

See Table 5.1 for population and lot numbers.

The smallest number of interviews were conducted in Ouro Preto municipal. Although this area had the highest rural population, it has the smallest number of farm lots and, therefore, was allocated the smallest number of interviews. Ouro Preto municipal has 440 lots, comprising of 7.9 per cent of the total lots in the city. Therefore 7.9 per cent or 12 of the total 150 goal interviews were conducted in this area. The remaining number of interviews were calculated in the same manner (Table 5.2)

There are two reasons why the Ouro Preto municipal has the highest rural population and the smallest number of lots. First, since this was the first area to be settled, the lots given away by the government were, on average, 100 hectares, while the lots given away and sold by the government in later years were smaller. Ouro Preto, therefore, has more people, on average, living on each lot. Second, the rural population was overestimated by the Brazilian statistical agency (Editora Turística e Estatística). In the early settlement years of the 1980s the rural population represented 11 per cent of the total population, but it is likely that this percentage has changed since the 1980s. However, it is this percentage that is used to estimate the 1994 rural population (Editora Turística, 1984). This percentage has most likely decreased over the ten-year settlement of the area as the central urban area has developed and

Table 5.1 Population Figures: Ouro Preto do Oeste Population 1994

Municipal	Rural	Urban	Total	Number of Lots	Average* Size of Lot (hectares)	Year of* Settlement
Ouro Preto	4462	40 156	44 618	440	101.53	1970
Valé do Pariso	1395	9097	10 492	1239	92.96	1976
Nova União	1492	9340	10 772	991	80.90	1978
Mirante da Serra	1360	8870	10 230	703	44.10	1980
Urupá	1763	11 001	12 765	1362	33.39	1980
Teixeropolis	1205	7860	9065	834	85.20	1976
Total	**11 617**	**86 324**	**97 941**	**5569**	**70.64**	

* based on the survey that was conducted as part of this research.
Source: 'Estado de Rondônia: Rodoviário, Político, e Estatístico: Nova Divisão Política', (Map, 1994) Editora Turística e Estatística LTDa, Gioāna, Goiás.

Table 5.2 Lot Statistics: *Agricultural Lots in Ouro Preto do Oeste 1996–1997*

Municipal	Fraction of Lots in Municipal	Calculated # of Interviews	Goal # of Interviews	# of Lots Between Interviews	Actual # of Interviews
Ouro Preto	0.0790	11.85	12	36.66	15
Valé do Pariso	0.2225	33.38	33	37.55	35
Nova União	0.1780	26.70	27	36.70	36
Mirante da Serra	0.1262	18.93	19	37.00	24
Urupá	0.2446	36.69	36	37.83	36
Teixeropolis	0.1498	22.47	23	36.26	25
Total	**1.0001**	**150.015**	**150**	–	**171**

Source: 'Município de Ouro Preto do Oeste' (Map, 1984) Ministério da Agricultura, Instituto Nacional de Colonização e Reforma Agrária, Coordenadoria Especial do Estado de Rondônia.

other, more rural counties have expanded around the city center. The overestimation of rural population is another reason that the number of lots is used for the stratified random sample rather than the rural population figures.

The distance between interviews, in terms of lots, was determined in order to represent the areas equally and to obtain spatial variation in topography, soil type and other spatial characteristics. These six areas of the city are similar in topography and soil type and have been settled according to soil production. The best soils were settled first, leaving the lower quality soils to later farm establishments. Again referring to the Ouro Preto municipal, since the goal number of interviews, 12, is determined to represent 440 lots equally in terms of space, every 37th house was chosen. However, the number of lots between interviews determined for each area were difficult to maintain since on average 30–50 per cent of the lots are uninhabited. Instead each 10–20 lots were interviewed, depending upon the number of uninhabited lots in the municipal.

If the house was unoccupied at the time of the interview or the owner of the lot was not available (or a relative of the owner), the next house on the same side of the road was interviewed, and if the same occurred at the next house, the next house was interviewed on the same side of the road. This occurrence was generally uncommon since most of the farmers stay close to, or on, their lots during the time of year the surveys were conducted (September–December), which is the end of the burning period and the beginning of the planting season.

There was only one refusal to participate in the survey. This was in the Mirante da Serra area. The high response rate is attributed to two factors: first, culturally, Brazilians are generally very hospitable to strangers and would go out of their way to make anyone comfortable; second, because of the lack of contact with people, the farmers were generally happy to see and talk to outsiders. Most of the farmers and their families do not leave their farms on a daily basis (especially during the rainy season) due to a lack of transportation, long distances to city centers and very poor road conditions.

The target number of interviews was reached and surpassed in five of the six municipals (Table 5.2). A few extra interviews were conducted, when possible, in case some of the interviews had to be dropped.

The majority of the 25 APA interviews were completed at the labor union/association headquarters when different members arrived for meetings or brought in their harvest of honey. Interviewing these members at the headquarters could possibly lead to a bias since the other interviews were conducted on the farm lots. Many of the APA members were visited during the five-month interview period to learn about and observe the intercropping systems. During these visits questionnaires were not administered, but these farmers were interviewed at APA's headquarters after these visits when other lots were difficult to visit. These visits minimized the possible bias that could result from conducting the interviews off the lot. These interviews were conducted at the labor union headquarters during the month of January, the height of the rainy season, when travel to the farm lots is difficult and often impossible.

5.4 THE SURVEY

The survey questions consisted of inquiries about the individual farmer, the family, characteristics of the lot, harvest, other forms of income in addition to farming, knowledge of sustainable agriculture, use of sustainable agriculture and major influences determining farming techniques. This information is applied to the estimation of the probability that a farmer will adopt sustainable agriculture or become a member of the organization APA, which supports the sustainable use of the forest. It is likely that these estimations rely greatly on farm and farmer characteristics.

The questions concerning individual farmers included inquiries about age, education level and farming experience. Family questions included information about family size, age of the family members, education levels, farming equipment owned, other types of equipment and number of farm animals (including cattle, pigs, chickens, sheep and goats). Characteristics like these generally influence the decisions that farmers and households make. For example, as the level of education of the household increases, it is expected that the knowledge of sustainable agriculture and groups like APA would increase, and therefore increase the probability of adopting sustainable agriculture or becoming a member of APA. It is expected that age, education, equipment and durable goods have a positive relation with the probability of adopting sustainable agriculture.

Lot characteristics that were collected include the size of the lot and the number of hectares of pasture, primary forest, secondary forest, agriculture and intercropping. It is expected that the division of the lot between forest, agriculture and pasture will affect the dependent variable: the probability of adopting sustainable agriculture. It is expected that those farmers who practice sustainable agriculture have larger areas of virgin forest since they do not have to cut and burn the forest to plant crops. However, a conflicting factor that may also increase the probability of adoption is a small amount of virgin forest. Those farmers who hold relatively small amounts of virgin forest may need to find an alternative way of farming as the amount of land that can be converted to annual crops approaches zero. This can support the effect that necessity can play in the adoption decision. It is expected that the average number of hectares burned per year, will decrease as the probability of adoption increases, since the farmers who use sustainable agriculture are expected to burn less on their lots.

Questions about agriculture income and income unrelated to agriculture were also asked. Income unrelated to agriculture includes income from social security and jobs such as the selling of perfume or crafts or any other service unrelated to farming. Questions about income derived from agriculture provided information about the harvest of all perennial and annual crops, milk harvest and meat harvest, the amount of each item that was sold, and at what price. It is expected that as income increases the probability of adopting sustainable agriculture will increase as well. Those farmers with higher incomes have a lower risk, which is associated with the adoption of new land uses. If the new land use fails, those farmers with higher incomes have greater income reserves to rely on.

The farmers' knowledge of any type of sustainable agriculture was recorded for both groups of farmers and, if sustainable agriculture was used, in-depth questions were asked as to the extent, the number of years of use, income generated, and income and crop harvest before the use of sustainable agriculture. This information is used to determine if income is greater for those farmers who have adopted sustainable agriculture.

The last section of the survey attempts to determine the major factors that could affect the agricultural practices of the farmer. Questions were asked to determine the extent of the effect that family, neighbors, free seedlings, free seed, bank loans, valuable wood and soil type have on the farmers' choice of agricultural techniques.

5.5 DATA OVERVIEW

Data collected in these interviews are used in the modeling and analysis chapter, Chapter 6. The variables derived from the data collection and their abbreviations are presented below. A brief overview of the most pertinent variables – including farm and family characteristics, lot characteristics, wealth measurements, income and harvest – are presented in the following tables. These variables are divided among the two data groups, non-APA members and APA members, and are shown in terms of mean, standard deviation (in parenthesis) and percentage (where appropriate).

5.5.1 Variables

Farmer characteristics
AGEHH – age of the household head (in years)
EDUHH – education in years of the household head (in years)
FARMEXP – years farming of the household head
FARMYEAR – years farming on lot of the family
UNION – number of associations/unions the family participates in
ST – state of birth
SP – state of birth of parents
OURO – dummy variable pertaining to Ouro Preto municipal
VALE – dummy variable pertaining to Valé do Pariso municipal
NOVA – dummy variable pertaining to Nova União municipal
TEIX – dummy variable pertaining to Teixeropolis municipal
MIR – dummy variable pertaining to Mirante da Serra municipal
URUPA – dummy variable pertaining to Urupá municipal

Lot characteristics
SIZELOT – size of lot
HECAGR – number of hectares devoted to agriculture

PERAGR – percentage of lot devoted to agriculture
HECPAS – number of hectares devoted to pasture/second growth
PERPAS – percentage of lot devoted to pasture/second growth
HECFOR – number of hectares devoted to primary forest
PERFOR – percentage of lot devoted to primary forest
HECSA – number of hectares devoted to sustainable agriculture
DISTOPO – distance to city center

Family characteristics
FAMILY – number of family members living on the lot
MEN – men>=10 years of age
BOY – men<10 years of age
WOMEN – women>=10 years of age
GIRL – women<10 years of age
AVEAGEHH – average age of husband and wife (in years)
AVEDUHH – number of years of education (average of husband and wife)
AGEW – age of woman of household (in years)
EDUW – number of years of education of woman of household
YEAR – year of migration to Rondônia
GARDEN – garden size (in square meters)
number of the following items owned: horses, chickens, cattle, pigs, televisions, sheep, satellite dish, goats, tractors, equipment for use with animals, cars, trucks, motorcycles, wheel barrows, telephones, refrigerators, bicycles, other lots, other houses, city houses, workers, chain saws, other types of equipment

Burning characteristics
BURNDAY – number of days of work necessary to burn (1996)
BURNP – number of people working during burning (1996)
EQP– equipment used to burn (1996)
AVEBURN – number of hectares burned (2 year average – 1995 and 1996)

Production characteristics
HARANN – 1995 harvest annuals in bags (rice, corn, beans, manioc)
HARPER – 1995 harvest perennials – measured in number of fruit with the exception of coffee, which is measured in bags (coffee, coco, banana, citrus, cupuaçú, pupunha, açaí, acerola, mahogany, other valuable wood, mango, coconut, jack, abacati)
Inputs: price and quantity [labor (days), fertilizer (bags), pesticides (bags), herbicides (bags)]

MILK – production in number of liters (rainy and dry season), in reais (1996)
TOTINPUT – total inputs used for agriculture in reais (1996)
MILKINC – total milk income in reais (1996)
MILKPRO – total milk production in reais (1996)

Sustainable agriculture
SEEDL – seedlings provided (1 = yes, 0 = no)
TOTINC – income after the use of sustainable agriculture in reais (1996)
USESA – use of sustainable agriculture (1 = yes, 0 = no).
KNOWSA – knowledge of sustainable agriculture
HECSA – hectares devoted to sustainable agriculture
YEARSA – year sustainable agriculture was first adopted
INCHO – other sustainable forms of income (beekeeping, fish-raising) in reais (1996)
HARBEF – harvest before the use of sustainable agriculture in the year preceding the adoption

Influences in farming techniques and choices (ranked from 1–3)
neighbors, association/union, free seed, free seedlings, bank loans, market for annuals, market for perennials, soil quality, standing precious wood, Ouro Preto farmers, parents

Income
MONEYINC – 1995–1996 income earned from agriculture [perennial+annuals+other forms+milk+income from sustainable agriculture (honey and fish)] – in reais (1996)
OTINC– other forms of income unrelated to agriculture in reais (1996)
FAMINC – total family income [perennial+annuals–inputs+other forms+milk+income from sustainable agriculture (honey and fish)] – in reais (1996)
PERCAP – income per capita – in reais (1996)
PERMILK – milk as percentage of income
PERHON – honey/fish as percentage of income
PERANN – annuals as percentage of income
PERPER – perennials as percentage of income
VEH – vehicles owned in reais (1996)
DURABLE – durable goods (vehicles+farming equipment+tractors+city houses) in reais (1996)
RES – resources (durable goods+cattle) – in reais (1996)
TOTINC – income after the use of sustainable agriculture – in reais (1996)

TOTINCB – income before the use of sustainable agriculture – in reais (year preceding the adoption)

5.5.2 Data Comparison

Table 5.3 presents farm and family characteristics for the two groups of data (APA and non-APA farmers) that were collected. Comparing the two groups of farmers, there is a significant difference in the age, farming experience and education levels between non-APA and APA farmers. APA members are on average younger, have less experience farming and have higher education levels. There is no significant difference in the number of family members on the lot between the two groups.

Table 5.4 compares the lot characteristics of the two groups of farmers in terms of lot size and the activity to which the land is devoted. APA members and non-APA members, on average, hold similar size lots, but the division of the land among agriculture, pasture, virgin forest and intercropping varies. APA members do not hold a significantly different number of hectares devoted to agriculture or pasture. However, APA members do hold a significantly greater number of hectares which are devoted to primary forest and intercropping. These comparisons support the theory that APA members have higher amounts of land devoted to the sustainable uses of the forest in comparison to the average farmer.

Table 5.5 presents the various ways in which wealth can be measured among the rural farmers. When we compare the two groups of farmers, APA members have less cattle, land and durable goods in terms of city houses, but none of these vary significantly from non-APA members. The only significant difference between the two groups in terms of wealth measurements is the difference in vehicle value. The value of vehicles is significantly higher for non-APA members. The value of the land is estimated on the basis of the resale value of the total lot for each municipal. The estimate of an average resale value was based on data on the average resale value per hectare for each municipal from the local labor union of rural farmers[1] and from farmers who had sold or bought lots in 1996. The value of the land is estimated on the basis of the value of land for each municipal and reflects the capitalized value of soil quality based upon annual crop productivity. The similarity in wealth between the two groups of farmers suggests that capital equipment and capital resources may not be required to adopt sustainable agriculture.

Table 5.6 provides information on family income and income distribution among different activities. APA farmers have, on average, a significantly higher dollar income in agriculture, thanks to a higher income from perennial

Table 5.3 Farm and Family Characteristics of Farms in Ouro Preto do Oeste

Farm and Family Characteristics*
comparison of the stratified random sample (non-APA members) and APA members

Characteristics	Non-APA Members (n=171)	APA Members (n=25)	t-statistic
Average Age of Household Heads (years)	46.20 (12.97)	33.62 (6.35)	7.77
Number of Years Farming of the Household Head	36.57 (14.89)	24.12 (7.39)	6.67
Number of Years on the Lot of the Household Head	17.18 (5.86)	16.48 (5.24)	17.18
Number of Years of School of the Household Heads	2.48 (2.46)	3.86 (2.10)	-2.9
Number of People on the Lot	8.42 (6.02)	9.89 (4.64)	-1.4

* The top number in each block represents the average and the number underneath in parenthesis represents the standard deviation associated with that mean.

crops (rather than annual crops). APA members also have a lower dollar income in terms of milk, but a higher average income from honey and fish. The income of APA member families is higher but not significantly different than non-APA members. Per capita income is, however, significantly lower for APA members. Further analysis is necessary in order to determine how the difference in family income affects agricultural practices. APA members do have a higher percentage of income arising from the activities that support sustainable uses of the forest, such as honey- and fish-harvesting and perennial crops, and a smaller dollar income (and smaller percentage of income) from

Table 5.4 Lot Characteristics of Farms in Ouro Preto do Oeste

Characteristic	Lot Characteristics* in hectares (1 hectare=2.47 acres)		
	Non-APA Member (n=171)	APA Members (n=25)	t-statistic
Size of Lot	70.64 (46.12)	74.45 (55.71)	-0.33
Agriculture	7.42 14.51% (6.54)	6.94 13.98% (5.87)	0.37
Pasture	46.45 61.54% (37.69)	33.46 43.58% (31.27)	1.89
Primary Forest	16.67 23.78% (18.88)	31.46 33.67% (34.09)	-2.12
Intercropping	0.11 0.17% (0.47)	2.59 8.78% (2.56)	-4.69
Average Burn/Year	6.70 (8.84)	2.025 (2.56)	5.51

* The top number in each block represents the average, the number underneath the first number represents the percentage of the lot devoted to each activity, and the third number, underneath, in parenthesis, represents the standard deviation associated with that mean.

the activities that are more destructive to the forest, such as the harvest of annual crops and milk harvest (which requires large plots of pasture). The income value in Table 5.6 does not include household consumed crops. For this information see Table 5.7, which includes a dollar value of all harvested crops, including those crops that are sold in local markets and consumed in the household.

Table 5.7 provides information on the total value of the agricultural harvest and the distribution of this value among different farm activities. Harvest value is the total harvest of agricultural crops in dollar terms. Crops

Sustainable Agriculture in Brazil

Table 5.5 Wealth of Farmers in Ouro Preto do Oeste

	Wealth Measurements* (in Brazilian Reais R$1= $US1)†		
Activity	Non-APA Members (n=171)	APA Members (n=25)	t-statistic
Cattle	16 250.00 (17 846.10)	12 280.00 (8763.29)	1.78
Durable Goods (farm equip., city houses, vehicles)	22 054.86 (24 003.88)	18 444.00 (13 821.01)	1.08
Farming equipment	2260.99 (26 601.99)	2958.00 (7333.34)	-0.43
Houses in the city	584.80 (2591.30)	2200.00 (5016.64)	-1.58
Vehicles	2959.07 (8145.26)	1000.00 (16005.23)	2.79
Land – total value of land, lot size is variable	26 225.61 (22 862.96)	25 468.00 (19 690.14)	0.18

* The top number in each block represents the average and the number underneath, in parenthesis, represents the standard deviation associated with that mean.
† This exchange rate represents the average exchange rate for 1996 according to the National Trade Data Bank and Economic Bulletin Board – products of Stat-USA, U.S. Department of Commerce (1997).

that were sold on the market are used in the estimation at their market values. If the crop was not sold by the family, the average price for the crop in 1996 is used in the estimation of the harvest value. The information in this table is similar to that of Table 5.6. APA farmers have, on average, a significantly higher harvest value in agriculture, due to a significantly higher harvest value from perennial crops (rather than annual crops). APA members also have a significantly lower harvest value in terms of milk, but a significantly higher harvest value from honey and fish.

The largest differences between the income values and harvest values are

Table 5.6. Income from Farm and Other Activities of Farmers in Ouro Preto do Oeste

Activity	**Family Income** (September 1995 – August 1996) (in Brazilian Reais R$1=$US1)†		
	Non-APA Members (n=171)	**APA Members** (n=25)	t-statistic
Agriculture (annuals+perennials- work and fixed costs)	2112.56 30.06% (4338.03)	3527.52 44.22% (2817.90)	-2.16
Annuals	496.01 12.30% (1017.52)	695.56 10.63% (828.38)	-1.09
Perennials	1616.55 21.67% (4206.50)	2831.96 35.49% (2797.34)	-1.90
Milk	2612.05 47.52% (3177.40)	1763.57 23.81% (1493.71)	2.20
Fish and Honey	9.06 0.08% (93.78)	1194.76 17.54% (964.03)	-6.15
Other (social security and any income unrelated to farming or agriculture)	1718.57 25.59% (4627.06)	182.96 16.64% (2058.18)	0.40
Total	6309.52 (7588.84)	6654.11 (4331.15)	-0.33
Total Per Capita	1055.40 (1790.63)	841.13 (496.37)	-7.08

* The top number in each block represents the average, the number underneath the first number represents the percentage of family income devoted to each activity and third number, underneath, in parenthesis, represents the standard deviation associated with that mean.

† This exchange rate represents the average exchange rate for 1996 according to the National Trade Data Bank and Economic Bulletin Board – products of Stat-USA, U.S. Department of Commerce (1997).

Table 5.7 Total Value of Harvest from Farm Activities of Farmers in Ouro Preto do Oeste

	Family Crop Harvest* (September 1995–August 1996) (in Brazilian Reais R$1=$US1)†		
Activity	**Non-APA Member** **(n=171)**	**APA Members** **(n=25)**	**t-statistic**
Agriculture (annuals+perennials- work and fixed costs)	4289.07 (8715.80)	8939.54 (5425.05)	-3.6
Annuals	2500.70 (4978.47)	3465.47 (3429.92)	-1.21
Perennials	2055.02 (5252.70)	4165.19 (3384.77)	-2.46
Milk	2652.51 (3137.62)	1841.18 (1391.52)	2.18
Fish and Honey	19.76 (156.17)	1578.82 (1030.18)	-7.40
Total (annuals+perennials+ milk+fish+honey)	7227.98 (9804.43)	11 050.66 (6001.90)	-2.7

* The top number in each block represents the average; the number in parenthesis represents the standard deviation associated with that mean.
† This exchange rate represents the average exchange rate for 1996 according to the National Trade Data Bank and Economic Bulletin Board – products of Stat-USA, U.S. Department of Commerce (1997).

the differences between total agriculture income (Table 5.6) and total agriculture value, and between perennial income (Table 5.7) and perennial harvest value. The amount by which milk income and milk harvest value differ is the same percentage for both the income and harvest values. The income for non-APA members is 1.5 times greater than the income for APA members, while the harvest value for non-APA members is also 1.5 times greater for non-APA members. The use of harvest values, therefore, does not make any relative changes in the income data. The differences between annual and fish and honey income and harvest values are also similar between the two groups. There is a large difference between income and harvest value of perennial crops for the two farmer groups. The income from perennial crops for APA members is 1.5 times greater than for non-APA members. The value of perennial harvest is however different. The harvest value of perennial crops is two times as large for APA members as for non-APA members. This suggests that the difference between the two groups of agriculture income and agriculture harvest value is due solely to the difference in perennial income and value.

Perennial fruits are harvested from the intercropping plots. These results suggest that APA farmers are growing more of the fruits that are supported by intercropping, but that the markets in which these farmers can sell these products are undeveloped or lacking since the products are not being sold to the extent that annual crops and milk are being sold.

The number of hectares burned on average, over a two-year period, and the percentage of the lot burned, based on the same two-year period, are plotted in the following two graphs to determine if the two data groups differ in terms of burning-behavior patterns. A two-year average is used because it was determined that the number of hectares burned can vary greatly from year to year for each household. Two years was, however, the longest period for which the farmers could answer accurately the questions that referred to their burning choices. Both of the figures (Figure 5.2 and Figure 5.3) plot the 'per cent of farmers' on the y axis. The per cent of farmers is plotted in order to account for the different sample sizes. A comparison of these two graphs supports the premise that the two groups of farmers differ in terms of burning behavior and reinforces the need for further analysis. There are significant differences between the two groups in the number of hectares that are burned (Table 5.4).

An analysis of these APA and non-APA members is made in Chapter 6 in order to determine the probability of adopting sustainable agriculture and whether the diffusion of sustainable agriculture can be increased with policy. The non-random sample, APA members, who support the sustainable use of the forest, provides a vital role in this analysis since the adoption of sustainable agriculture is very low among non-APA members. Less than 12 per cent of the stratified random sample uses intercropping and/or harvests honey. The non-random sample of relatively homogenous households allows for additional inferences to be made about land-use decisions. Variables such as income, lot size, the percentage of the lot burned, land value, the number of unions in which the family participates and the percentage of the lot that is primary forest are used in the regression analysis found in Chapter 6.

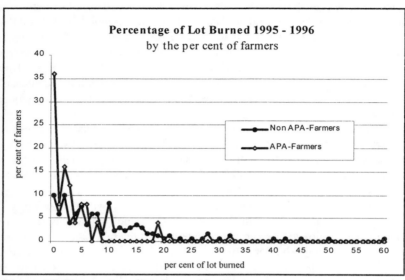

Figure 5.2 Percentage of Lot Burned by Farmer

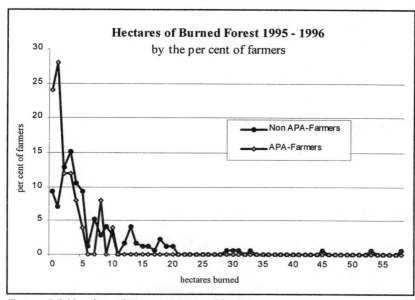

Figure 5.3 Number of Hectares Burned by Farmer

ENDNOTES

1. Sindicato dos Trabalhadores Rurais (STR), Ouro Preto do Oeste, Rondônia.

Chapter 6

The Economic Model

6.1 FARM CHOICES AND DECISIONS

The objective of this study is to test the role that different policy options can make on the farm family's decision to adopt sustainable agriculture. It has been asserted that the adoption of sustainable agriculture by farmers who use slash-and-burn agriculture can reduce deforestation rates considerably if adopted by enough farmers. One of the key issues that must be addressed is why the rate of adoption of sustainable agriculture is low in the tropical forests of Rondônia, Brazil. If it can be determined why the diffusion rate is low, policies that address these issues can help to increase the adoption rate.

The choice to use sustainable agriculture is not necessarily a discrete one; instead, a spectrum of alternatives exists. It is possible for a family to choose to convert a portion of its lots to sustainable agriculture and maintain annual crops, coffee and pasture on other portions of the land. This option reduces the risk that may be involved in the sole use of sustainable agriculture. Risk is involved with the adoption of sustainable agriculture because the farm family is not familiar with the agricultural technique or the markets where these products can be sold. There is no certainty that the agricultural products that are cultivated in the sustainable agricultural systems can be harvested and sold or provide the same profit range as annual crops. The markets for these products are not as formally developed as are the markets for annual crops and coffee.

Studies that have addressed local market failure in developing countries by the adoption of new techniques of agriculture have investigated a wide variety of technologies. A common theme throughout these papers is that the adoption of new techniques which increase welfare and/or income often occur

64

at low diffusion rates (Holden 1993, Kebede *et al.* 1990, Sands 1986, Opare 1977). Many farmers are either not willing to adopt the new technology or are constrained by capital resources and cannot adopt the potentially superior technology. This observation was also made in Ouro Preto do Oeste. Only 10 per cent of the farmers from a random sample used sustainable agriculture even though the technique was shown to improve the welfare and income of those farmers who used it.

One might expect that sustainable agriculture would be adopted in Ouro Preto do Oeste because it requires less average labor per year for the farmer compared to slash-and-burn agriculture[1] and also because the use of sustainable agriculture has the potential to increase the profits of the family. Another benefit of adopting sustainable agriculture is that it does not require that the family burn their land each year to plant crops. It was found, in a random sampling of the farmers, that they burned an average of three hectares per year. Although this amount is small compared to a size of a majority of the lots,[2] each hectare of burned forest means that there is less land available for future generations. Since property rights are now well established in the area, there is an incentive to maintain the production capability of the land since the property is an intergenerational investment. There are also costs which are incurred when sustainable agriculture is adopted. These costs include the initial set-up cost (although for many farmers this is zero in Ouro Preto do Oeste since many of the seedlings which are used in intercropping have been provided free of charge by local agencies) and the risk which is associated with the adoption of a new technology in terms of production and income capability. Since there are benefits to adopting sustainable agriculture, this study focuses on policies which increase these benefits and make them evident to the farmers in the area.

6.2 FAMILY UTILITY MODEL

The decision to adopt sustainable agriculture is treated as a dichotomous choice. The farm family decision is made on the basis of the utility derived from the chosen agricultural technique. The family chooses between using sustainable agriculture or slash-and-burn agriculture. This choice is not exclusive. When the decision is made to adopt sustainable agriculture the family may continue to cultivate annual crops.[3] This utility is compared to the utility that the farm derives from slash-and-burn agriculture. In the original conception of this study, the adoption of sustainable agriculture was to be treated as a pure discrete choice decision. The farm family used either slash-and-burn agriculture or sustainable agriculture. However, as the interviewing progressed it was found that a majority of farmers who adopted sustainable agriculture also maintained annual crops on a portion of their lots.[4] The choices that face the farmer are therefore represented in this analysis as choosing to use slash-and-burn agriculture or using sustainable agriculture exclusively or in combination with slash-and-burn agriculture. The farmer therefore chooses the combination of farming techniques that delivers the maximum utility. Since the adoption of sustainable agriculture is not a pure discrete choice decision, the extent of adoption is also estimated for those farmers who adopt sustainable agriculture.

A utility maximization model is developed according to the supposition that farm families choose a set of agricultural practices based on the resources that are available, the knowledge they possess and the constraints that limit these activities. The utility derived from the agricultural choices is not directly observable. The non-observable underlying utility function that ranks the preference of the farm family is derived from farm characteristics and characteristics of the specific technology.

The utility model chosen to represent the family decisions of the small-scale farmers follows the family utility–family budget constraint model of Killingsworth (1983). Family utility, based on the combined family income and leisure, is utilized in this study instead of individual utility, since this best represents small farm behavior in Amazonia. Given that the individual consumption within the family unit cannot be observed, the family utility is the best measurement of farm activity. Discussions with the families in Ouro Preto do Oeste indicated that families act and make decisions as a single unit. Management decisions on crop production, distribution, inputs and mixes, the timing of operations and work allocation are made by members of the household according to household goals and needs.

The family utility–family budget constraint model works similarly to the simple labor supply framework with one exception, which Ashenfelter and Heckman (1974) discuss. In this model the individual makes work decisions based on the effect of labor inputs of himself or herself and the labor inputs of all other family members. The most popular version of this model makes the assumption that these two substitution effects are reduced to a pure income effect (Killingsworth 1983). This same assumption, which reduces the demand equation to a simpler utility framework that behaves as a single consumer utility model, will be used in this analysis. For other studies that make this same assumption see Bourguignon (1989), Bigsten (1988), Bognanno *et al.* (1974) and Cohn *et al.* (1970).

6.3 THE CONCEPTUAL FRAMEWORK

Following Adesina and Zinnah (1993) and Rahm and Huffman (1984), the utility maximization of the family utility model is based on the non-observable underlying utility function, which ranks the preference of the *ith* family according to the technology chosen. The non-observable underlying utility function is represented by $U_{it}(M_i, T_i)$, where *t* represents the technology choice (*t*=1 when sustainable agriculture is used and *t*=2 when slash-and-burn agriculture is used exclusively). The utility is derived from the observable farm and family characteristics, *M* (such as lot size, percentage of the lot which is primary forest and the number of hectares of secondary growth), and from the observable technology characteristics, *T* (such as yield, income and the labor and leisure ratio), where the technology refers to the agricultural method of the family. The choice of technology therefore determines the farm yield, family income, and the labor and leisure that the family can afford. The choice of the technology makes this determination since the crops that are harvested and method of planting differ according to the farm technique. Although the utility is unobservable, the relation between the utility derived from a specific technology is a function of the vector of the observed farm and technology characteristics included in the utility measurement. Therefore, although the utility is unobservable and since its value cannot be calculated directly, the value may be implied by the farmer (firm-specific characteristics, *X* (e.g. agricultural technique, lot size and soil productivity). The family chooses between U_{i1} and U_{i2} depending upon which technology yields the greatest utility. The utility of the chosen technology is therefore estimated from the vector of observable farm and technology characteristics as follows:

$$U_{it} = \alpha_t F_i(X_i) + e_{it} \qquad t = 1,2; \quad i = 1,\ldots\ldots,n \qquad 6.1$$

where e_{it} is a disturbance term having zero mean.

Equation 1 is not restricted to be linear. The exact distribution of *F* depends on the distribution of the error term. The *ith* family will choose to use slash-and-burn agriculture if $U_{i1} < U_{i2}$, or if the latent variable $Y^* = U_{i2} - U_{i1} > 0$ and will choose sustainable agriculture when $U_{i1} > U_{i2}$, or if the non-observable latent variable $Y^* = U_{i1} - U_{i2} > 0$:

$$Y_i = \begin{cases} 1 & \text{if } U_{i1} > U_{i2} \text{ } \textit{sustainable agriculture is adopted} \\ 0 & \text{if } U_{i1} \le U_{i2} \text{ } \textit{slash-and-burn agriculture is adopted} \end{cases} \qquad 6.2$$

The probability that the farmer adopts sustainable agriculture or that Y_i equals one is a function of the independent variables:

$$
\begin{aligned}
P_i \quad &= \quad Pr(Y_i=1) = Pr(U_{i1}>U_{i2}) \\
&= \quad Pr\,[\,\alpha_1 F_i(X_i) + e_{i1} > \alpha_2 F_i(X_i) + e_{i2}] \\
&= \quad Pr\,[(e_{i1} - e_{i2}) > F_i(X_i)(\alpha_2 - \alpha_1)] \qquad\qquad 6.3 \\
&= \quad Pr\,[\mu_i > - F_i(X_i)\,\beta] \\
&= \quad F_i(X_i\,\beta)
\end{aligned}
$$

where X is an n x k matrix of explanatory variables, and β is a k x 1 vector of coefficients to be estimated.

$$
U_{it} = \alpha_t F_i(M_i, T_i) + e_{it} \qquad t = 1,2; \quad i = 1,\ldots\ldots,n \qquad 6.4
$$

where e_{it} is a disturbance term having zero mean.

The probability that the i*th* family adopts sustainable agriculture is the probability that the utility gained from slash-and-burn agriculture is less than that gained from sustainable agriculture or the cumulation distribution function of F for μ_i^5 evaluated at $X_i\beta$. If u_i is normal, F will have a cumulative normal distribution, and if u_i is uniform then F is triangular (Adesina and Zinnah 1993). For the purpose of this analysis, u_i is assumed to be normal, making the estimation of the probability possible using a probit model (see Appendix D for a discussion of discrete choice models).

The second stage of the model involves the estimation of the extent of adoption once the decision is made to adopt sustainable agriculture by the i*th* family. Evaluating equation 6.4, the functional form of F is specified with a Heckman model where u_i is an independently normal distributed error term with a zero mean and constant variance σ^2.

$$
\begin{aligned}
Y_i = X_i\,\beta \quad &\text{if} \quad i* = X_{i2}\beta_2 + \mu_i > H \\
Y_i = 0 \quad &\text{if} \quad i* = X_{i2}\beta_2 + \mu_i \le H
\end{aligned}
\qquad 6.5
$$

Y_i is the probability of adopting sustainable agriculture, $i*$ is a non-observable latent variable, H is a non-observable threshold value, and X_{i2} are the independent variables used to explain the extent of adoption decision. Unlike the tobit model, those variables which determine adoption and those which determine extent of adoption differ.

After estimating the probability and extent of adoption, estimation of the marginal effects are necessary since probit model coefficients do not represent the expected change in the dependent variable given a one-unit change in the explanatory variable. The marginal effect of a variable X_j on the probability of adopting the new technology is:

$$\frac{\partial P_i}{\partial X_{it}} = f(X_i \beta) \beta_t \qquad 6.6$$

where $f(X_i \beta)$ is the marginal probability density function of u_i (Adesina and Zinnah 1993). The direction of the marginal effect is, therefore, determined by the sign of β_t. The coefficient β_t represents the coefficient differences ($\alpha_{2t} - \alpha_{1t}$). Thus β_t is expected to have a positive (negative) sign if the coefficient on the slash-and-burn coefficient is larger than (less than, equal to) the coefficient on the sustainable agriculture coefficient.

Farm and farmer characteristics such as lot size, average education of the household heads (average of the husband and wife) and average age of the household heads are included in X. It is expected that the probability will increase with lot size and the number of years of education, and decrease with age. The probability is expected to increase with lot size and the number of years of education because, as lot size and education increase, so do human capital and monetary resources, and as capital increases it is more likely that a superior technology will be adopted. Also expected is that as age increases, it is less likely that the farmer will be willing to change habits and therefore adopt the new technology (Adesina and Zinnah 1993, D'Souza *et al.* 1993, Hussain *et al.* 1994, Lin 1991, Norris and Batie 1987). Also included are capital variables such as labor available to the family since the labor requirement varies between the two technology choices. It is expected that the probability of adoption will increase as these variables increase.

The technology characteristics included in X included income per capita and the labor–leisure ratio. Also included in the technology characteristics are variables that indirectly affect the probability of adoption through information. Information about the potential gains in income and leisure from the use of sustainable agriculture can influence the rate of adoption. Union membership, indicated by the number of unions that the family participates in, is therefore included in X.

The economic model suggests that farm and family characteristics such as educational level, information and union membership may be important determinants of farmer choice. It is important to know if these variables play a significant role in this choice because these factors can clearly be influenced by policy. Other variables such as age, farm size and labor availability may also play large roles, but these, cannot however, be influenced through policy. If these policy variables do play a significant role in this analysis, the policy recommendations can lead to important results for other parts of Rondônia state, where similar programs have been initiated.

6.4 DATA LIMITATIONS

The data used in the analysis of farmer behavior in Ouro Preto do Oeste, Rondônia, were collected over a five-month period and are divided among two different groups. The first group consists of 171 observations that were collected according to stratified random sampling of farm families (including all people living on the lot). The second group consists of 25 observations that were collected from farmers who participate in the local farmer union, the Association of Alternative Producers (APA). This association supports the use of sustainable agriculture.

The degree of transferability of the results of the analysis of these farmers may be limited since the data were collected in a single settlement in a single year. Obviously it would be preferable to gather a broad sample of farmers from different settled areas in the states of Rondônia, Mato Grosso, Pará, Acre and other periphery states. Another limitation of these data is they do not measure changes over time, since they have been collected in one time period. A panel data set that tracks the choices of the farmers over time could provide a great deal of information about farm choice. Nevertheless, the results still provide insight into the farm choice and the diffusion of sustainable agriculture in Ouro Preto do Oeste and can help to provide preliminary policy recommendations that aim to increase the rate of adoption of sustainable agriculture across Amazonian states. Even if these policy actions are not applicable to a wide range of farm establishments, we may still learn practical lessons about the adoption of sustainable agriculture through this analysis. Other land use policies and manpower development policies do exist, although tailoring them to this problem may be difficult.

6.5 THE EMPIRICAL MODEL

The probability of adopting sustainable agriculture and the extent (or intensity) of adoption is estimated using a two-stage Heckman model. This model was chosen over both the probit and logit models, which only estimate the probability of adoption, because the extent of adoption is pertinent to policy implications. Both the Heckman (Heckman 1979) and tobit (Tobin 1958) models estimate adoption and the extent of adoption based on the concept of a threshold value of the dependent variable (Pindyck and Rubinfeld 1981).[6] The extent of adoption hinges on the adoption decision. Therefore, failure to model the adoption decision can lead to a selection bias. For example, one may find that policy influences the extent of adoption only. It is possible to ignore the probit stage, however such an estimation suffers from a selection bias. Given that the decision to adopt sustainable agriculture is characterized by a dichotomous choice, there is a point where the extent of adoption cannot be measured. However, when the threshold value, which is determined by a vector of observable explanatory variables, is exceeded, adoption occurs and the second decision on the intensity of use is taken. See Appendix D for a detailed discussion of these models.

The Heckman model (Heckman 1979) allows for different variables that determine the adoption choice and then the effort allocated when adoption is chosen. The two-equation procedure involves the estimation of a probit model of the adoption decision, calculation of the sample selection control function, and incorporation of that control function (the inverse Mills ratio) into the model of effort, which is estimated with ordinary least squares (OLS).[7]

A two-stage Heckman model[8] is estimated rather than the more commonly used two-stage tobit model [9] since it is expected that different explanatory variables influence the decision to adopt sustainable agriculture and the extent to which it is adopted. The tobit model makes the assumption that the variables that determine both the adoption and extent of adoption are identical. For example, the adoption of sustainable agriculture can be influenced greatly by the knowledge that this method of agriculture exists, but it is unlikely that this knowledge (represented as dummy variable) influences the extent of adoption. Also, in reverse, variables that may influence the extent of adoption may include the number of years that sustainable agriculture is used and the level of labor available to the family. It is unlikely that the knowledge that sustainable agriculture exists will influence the extent of adoption and impossible that the number of years of use will influence the decision to adopt sustainable agriculture.

This analysis is an addition to the literature on adoption behavior because a model that is rarely employed in technology adoption, a Heckman model of estimation, is used. Most studies estimate the extent of adoption using the tobit model, since the elasticities, evaluated at the means, can be decomposed into the elasticity of adoption and the elasticity of effort given that adoption occurs (Adesina and Zinnah 1993, Gould *et al.* 1989, Norris and Batie 1987).[10] These decomposed elasticities may be interesting for the interpretation of policy since they provide information and allow the elasticity of adoption and the elasticity of effort, but it is asserted here that the assumption that is necessary to calculate these elasticities is not relevant to this research.

Three series of estimations are performed in order to examine the probability of adopting sustainable agriculture and the extent of adoption (Figure 6.1). The results from the models are compared in order to determine the role that APA membership can make in the adoption of sustainable agriculture and the variables that play a role in the adoption of sustainable agriculture for non-APA members. The first series of equations that are analyzed are the probability of adopting sustainable agriculture and the extent to which it is adopted using the stratified random sample (non-APA observations). The extent of adoption is measured in two different equations, first in terms of the number of hectares devoted to intercropping and second in terms of liters of honey harvested.[11]

The second series of equations that are analyzed are the extent of adoption for APA members. Since all APA members participate in sustainable agriculture, a probability of adoption equation was not estimated. The extent of adoption is estimated using the number of hectares devoted to intercropping and the liters of honey harvested, as above.

The third series of equations estimated include the observations from both data groups. The non-APA and APA members are weighted according to the ratio of the population probability and the sample probability that is applicable

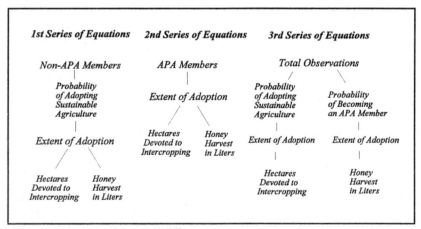

Figure 6.1 Series of Estimated Equations

to each of the groups of data: non-APA and APA. This weight accounts for the sampling variation in non-APA observations that may exist if the stratified random sample contains a sampling bias and for the selectivity bias of the APA member sample (see Section 6.7 for a discussion of the calculation of the weights and the modeling based on these weights). Two estimations are made using the weighted combined sample. One is the probability of adopting sustainable agriculture and the extent, measured in both number of hectares, and liters of honey. The second estimations are the probability of becoming an APA member and the extent of adoption once a farmer becomes an APA member. The probability of becoming an APA member is compared to the model estimated for the extent of adoption using the 25 APA observations to determine if a selectivity bias exists.

6.6 FIRST SERIES OF EQUATIONS – THE PROBABILITY OF ADOPTION

The first series of equations are estimated on the basis of the stratified random sample[12] to determine the probability of adoption and the extent of adoption once the adoption choice is made. Of the 171[13] farmers who make up this sample, 17 use sustainable agriculture, which represents 9.9 per cent of the sample. Although the adoption rate is low, this sample can provide insight into the adoption decision and, in turn, help to determine why the adoption rate is low.

The two-stage model is run using dummy variables for each of the municipals in the city of Ouro Preto do Oeste, with the exception of Ouro Preto, the municipal located in the city center.[14] In addition the model estimates the probability of adoption by using variables for education of the household heads,[15] age of the household heads, knowledge of sustainable agriculture, the number of unions the family participates in, the number of years that the family has resided on the farm, and per capita income. The variables included in this model and the following estimates are defined in Table 6.1.

There have been many attempts to derive goodness of fit measures for discrete choice models (Greene 1997). An analogue to the R-squared used in conventional OLS regression models has been developed and is the most popular of the fit measures for binary choice models. The pseudo-R-squared or the likelihood ratio index (LRI) was developed as a fit measure for probit, logit and tobit models (Gujarati 1995). This measure has an intuitive appeal in that it is bounded by 0 and 1. There is no way for the LRI to equal 1, but it can come close (Greene 1997). Values between 0 and 1, however, have no

Table 6.1 Variable Definitions

Variable	Definition
OPO	dummy variable for the municipal Ouro Preto, opo=1 if the family resides in Ouro Preto, opo=0 if the family does not reside in Ouro Preto.
NOVA	dummy variable for the municipal Nova União, nova=1 if the family resides in Nova União, nova=0 if the family does not reside in Nova União.
VALE	dummy variable for the municipal Valé do Pariso, vale=1 if the family resides in Valé do Pariso, vale=0 if the family does not reside in Valé do Pariso
TEIX	dummy variable for the municipal Teixeropolis, teix=1 if the family resides in Teixeropolis, teix=0 if the family does not reside in Teixeropolis.
MIR	dummy variable for the municipal Mirante da Serra, mir=1 if the family resides in Mirante, mir=0 if the family does not reside in Mirante.
URUPA	dummy variable for the municipal Urupá, urupa=1 if the family resides in Urupá, urupa=0 if the family does not reside in Urupá.
AVEDUHH	average education of the household heads in years
AVEDU2	average education of the household heads squared in years
AVEAGEHH	average age of the household heads in years
AVEAGE2	average age of the household heads squared in years
UNION	number of unions in which the household participates
UNIONAPA	number of unions that a household participates in excluding APA
KNOWSA	knowledge of sustainable agriculture, knowsa=1 if the family knows of or has heard about sustainable agriculture and knowsa=0 if the family does not know of or heard about sustainable agriculture
YEARONFM	number of years on the farm lot
YEARSSA	number of years that the family has used sustainable agriculture
PERBURN	percentage of the lot which has been burned
MONEYINC	total income gained in 1996, 1996 reais
PERCAP	total income gained in 1996, 1996 reais per capita
NOHONCAP	income from sources other than honey in 1996, total family income per capita – honey income per capita in 1996 reais
PERFOR	percentage of the lot which is virgin forest
PERPAS	percentage of the lot which is pasture or secondary growth forest
ADULTMAL	number of males over ten years of age

natural interpretation (Greene 1997). This value will be presented in the following models, since it is commonly referred to, but it will not be discussed further since it is lacking natural interpretation. The LRI index for this model is 0.33. The coefficient estimates for this model are presented in Table 6.2.

The model that estimates the probability of adoption predicts 90 per cent of the equations correctly. The model predicts correctly 92 per cent of the households that do not adopt sustainable agriculture and 64 per cent of the households that adopt sustainable agriculture (Table 6.2). The R-squared is 0.34. A low R-squared is to be expected, since an upper bound R-squared for binary choice models is about 0.33 (Pindyck and Rubinfeld 1991).[16] It is for this reason that R-squared values are not a relevant method of measuring the fit of probit models. The R-squared values will, therefore, not be reported in the remaining probit estimation. Instead, the likelihood ratio, a Chi-squared statistic test, is interpreted to determine fit of the probit models.

Both the probit and tobit models use maximum likelihood methods (MLE) to estimate the coefficients of the model. The regression coefficients are asymptotically efficient, unbiased and normally distributed (with large samples). The ratio of estimated coefficients and its standard error approximates to a normal distribution. Therefore, an analogue of the t-test is applied as a test for significance. In the MLE, a log likelihood test replaces the usual F test of OLS regression models to evaluate the significance of all coefficients. The log likelihood ratio test follows a Chi-square distribution with k degrees of freedom. For this model, the Chi-squared value of 43.83 at 13 degrees of freedom is significant at the 1 per cent level implying that the independent factors, taken together, influence the adoption of sustainable agriculture.

The most significant variable in determining the probability of adopting sustainable agriculture is the knowledge that sustainable agriculture exists. As this knowledge increases, so does the rate of adoption. A positive and significant sign on the union variable indicates that as the number of farm-related unions a family participates in increases, so does the likelihood of adoption of sustainable agriculture. Also affecting the adoption positively is the number of years on the farm (YEARONFM). The number of years that a family has resided on the lot can influence the adoption of sustainable agriculture in two ways. One is that as the years increase, so does the access to seedlings and opportunity to plant them since the programs providing the seedlings, free of charge, began about ten years ago and are continuing. The seedlings have been provided free of charge by local unions or government-supported agencies. Also, as the number of years increase, so does the permanency of residency of the family and future generations and,

Table 6.2 Coefficient Estimates for the Probability of Adopting Sustainable Agriculture and the Extent of Adoption

	Probability of Adopting Sustainable Agriculture (n=170)		Extent of Adoption (Hectares) (n=21)	Extent of Adoption (Honey) (n=21)
Variable	Coefficient	Marginal Impacts[a]	Coefficient	Coefficient
Constant	-5.8489 ** (-1.989)	-0.716	-0.181 (-0.117)	707.04 (0.285)
NOVA	-0.43306 (-0.784)	-0.053	-1.9212 ** (-3.844)	-1008.9 (-1.253)
VALE	-0.34414 (-0.609)	-0.042	-0.82478 * (-2.738)	-11.349 (-0.024)
TEIX	0.16804 (0.301)	0.020	-0.22926 (-0.971)	-358.64 (-0.936)
MIR	0.04618 (0.072)	0.006	0.26861 (0.742)	-420.94 (-0.725)
URUPA	-0.45916 (-0.756)	-0.56	-0.67460 ** (-1.845)	-105.93 (-0.181)
AVEDUHH	0.31918 (1.551)	0.039	0.14787 (0.803)	-115.73 (-0.395)
AVEDU2	-0.241E-01 (-1.107)	0.003	-0.314E-01 (-1.452)	8.5135 (0.246)
AVEAGEHH	0.099172 (0.949)	0.012	0.439E-02 (0.091)	-47.245 (-0.609)
AVEAGE2	-0.786E-03 (-0.773)	-0.000	0.2869E-04 (0.066)	0.38425 (0.550)
UNION	0.40043 ** (2.167)	0.049	0.3690 *** (2.660)	293.11 (1.342)
KNOWSA	1.7685 *** (4.092)	0.216		-0.22519 (-0.028)
YEARONFM	0.046020 * (1.559)	0.006	0.386E-02 (0.274)	-12.011 (-0.529)
YEARSSA			0.2035 *** (4.454)	64.918 (0.902)
PERCAP	0.228E-04 (0.338)	0.000	-0.111E-03 (-1.439)	
PERFOR			0.13E-01 * (2.701)	

Table 6.2 Continued

Variable	Probability of Adopting Sustainable Agriculture (n=170)		Extent of Adoption (Hectares) (n=21)	Extent of Adoption (Honey) (n=21)
	Coefficient	Marginal Impacts[a]	Coefficient	Coefficient
ADULTMAL			-0.10731 ***	1.6263
			(-3.334)	(0.032)
NOHONCAP				0.017162
				(0.141)
LAMBDA			0.36846	609.68
			(1.276)	(1.314)

Probability Model:			Extent of Adoption (Honey):	
Likelihood ratio test	42.83***		F statistic	1.21
Pseudo R²	0.34		R-squared	0.82
Number of Correct predictions	153 (90%)		adjusted R-squared	0.14
Extent of Adoption (Hectares):				
F statistic	13.76 ***			
R-squared	0.98			
adjusted R-squared	0.91			

***, **, * indicate significance at the 1%, 5% and 10% levels, respectively.
[a]Marginal impacts are the partial derivatives of $E[y] = \Phi[*]$ with respect to the vector of characteristics. They are computed at the means of the independent variables.

consequently, the need to maintain a land use that is sustainable. It is likely that as these families mature, the families may be concerned about land production for future generations (since property rights are secure in the city) and also may simply understand their land better, enhancing their ability and understanding of sustainability.

Unions, networking and the dissemination of information appear to greatly increase the probability of adoption, but at the same time neighboring effects appear to be insignificant. One might expect that neighbors might be responsible for the dissemination of the information, but the model does not support this since none of the dummy variables that represent municipals are significant at the 10 per cent level. Instead, unions appear to serve as the main source of information and the dissemination of information on sustainable agriculture. Knowledge of sustainable agriculture, which is often provided by unions and government-sponsored programs, greatly increases adoption. These results suggest that increasing this information, or increasing the rate at which information is spread, could increase adoption considerably.

Per capita income is not significant. Economic theory would predict that per capita income would play a role in the adoption of new technologies, but it has been shown in various studies of developing countries that income does not play a large role; information, education and/or resources are better determinants of the adoption of new technologies in developing countries (Hussain *et al.* 1994, Adesina and Zinnah 1993, Lin 1991, Kebede *et al.* 1990, Shakya and Flinn 1985). For this reason, these studies do not include income measurements in their models. This theory is also supported by the results in this analysis. These results also suggest that families having relatively higher educations and a greater number of years on their farm lot are more likely to adopt sustainable agriculture.

The coefficient estimates provided in Table 6.2 are not comparable with one another because the magnitudes of the coefficients depend on the unit of measurement and because they are expressed as indices. Unlike OLS coefficients, probit coefficients do not represent the expected change in the dependent variable, given a one-unit change in the explanatory variable. Thus, they have little interpretive value unless transformed into marginal values. The marginal impacts of changes in the independent variable are estimated for the significant independent variables. The interpretation of these marginal impacts are dependent upon the units of measurement of the independent variables (Gould *et al.* 1989). For example, these marginal impacts show that as education increases by one year, the probability of adopted sustainable agriculture increases by 3.9 per cent. The direction of the per centage change is as would be expected. As the number of years of education increases, the probability of adopting superior technologies also increases.

The variables that have important policy implications are knowledge of sustainable agriculture (KNOWSA) and union membership (UNION). The marginal impacts show that when farmers become informed about sustainable agriculture the probability of adoption is expected to increase by 22 per cent. These marginal impacts also show that when a family participates in one additional union, the probability of adoption increases by 4.9 per cent.

6.6.1 First Series of Equations – Extent of Adoption in Hectares Devoted to Sustainable Agriculture

The extent of adoption is estimated in terms of (1) the number of hectares devoted to sustainable agriculture or intercropping and (2) the number of liters of honey harvested in one year (1996). They are both estimated using the Heckman selection bias correction. The extent in terms of hectares devoted to intercropping is estimated first.

The extent of adoption is first estimated in terms of the number of hectares devoted to sustainable agriculture. The model is estimated in order to determine the extent of adoption of sustainable agriculture conditional on the adoption decision. This model includes dummy variables for each of the five municipals (as in the first stage). In addition to variables for age, education, union membership, per capita income and the number of years that a family has lived on the lot, the number of years that a family has practiced sustainable agriculture is included in the estimation. Also, this model includes variables that represent family capital, the number of adult males (ADULTMAL) and the percentage of the lot that is primary forest (PERFOR), indicating how much of the lot may be preserved if sustainable agriculture is used.

The F statistical value is significant at the 1 per cent level, suggesting that the overall significance of the model is high. The R-squared value is 0.98 and the adjusted R-squared value is 0.91, suggesting a high degree of fit for the model (Table 6.2).

Neighboring effects appear to play a more dominant role in the number of hectares devoted or converted to intercropping, particularly in the municipals of Nova União, Urupá and Valé do Pariso. The negative coefficients signify that compared to Ouro Preto these areas devote less land to intercropping. This result is particularly interesting in terms of the characteristics of Urupá and Valé do Pariso. Urupá was the last municipal to be settled in the area. The soil is relatively fertile and, as a result of this and the year of the settlement, the lots are significantly smaller than the other five municipals in Ouro Preto do Oeste. The largest lots in Urupá are 50 hectares and the smallest, 10 hectares. In comparison, the majority of the lots in the five remaining municipals are 100 hectares. The residents of Urupá therefore devote less land to sustainable agriculture because they have less land. The residents in Valé do Pariso also have less land to devote to sustainable agriculture because of the condition of the land. The majority of the lots in Valé do Pariso have boulders throughout, making much of the land impossible to farm.

Two variables that are representative of the resources that may be obtained from outside organizations are the number of unions that the family participates in (UNION) and the number of years that the family has used sustainable agriculture (YEARSSA). UNION and YEARSSA are positive and significant at the 1 per cent level, indicating that, as the number of unions in which the family participates increases, and as the number of years that the family has used sustainable agriculture also increases, so does the extent of adoption. Since the seedlings used in the intercropping plots are provided free of charge by the local unions and local governmental programs, an increase in the number of unions in which a family participates or in the number of years that the family uses sustainable agriculture can also increase the availability of these seedlings. As the seedlings become more available, the extent of adoption increases.

The coefficient on one of the measurements of capital resources, adult males (ADULTMAL), is negative and significant at 1 per cent. This negative relation suggests that the families who adopt intercropping do so to take advantage of the less intense labor technique. As stated earlier, intercropping requires less labor per year compared to the slash-and-burn method of farming annual crops. It is those families that do not have extensive labor resources (those that aim to reduce work load) that are more likely to adopt sustainable agriculture. Those families that do not have a large amount of labor on hand (adult males) are choosing to convert more of their land to intercropping compared with those farmers having greater labor and savings options.

The percentage of the lot remaining as undisturbed virgin forest (PERFOR) has a positive coefficient and is significant, suggesting that the farmers who have a greater percentage of standing forest may be more aware or better educated about the value of the standing forest. These farmers are, therefore, likely to devote more land to this potentially sustainable method of farming and potentially to preserve more land in the future.

6.6.2 First Series of Equations – Extent of Adoption in Liters of Honey Harvested

The extent of adoption is secondly estimated in terms of the number of liters of honey harvested. This second model is based on the same Heckman correction technique used in the estimation of the extent of adoption in terms of the hectares devoted to intercropping. The estimation of the extent of adoption in liters of honey does not appear to be a strong predictor of household decisions. The F statistical value is not significant at the 5 per cent level, suggesting that the overall significance of the model is low. The R-squared value is 0.83 while the adjusted R-squared is 0.14, suggesting that there are factors that have not been represented in the model which may explain the extent of adoption.

Surprisingly, none of the variables are significant. It is expected that the income that is not derived from honey, but earned outside the farm and from other farm products, would have an influence on the amount of honey harvested. The per capita income that is not derived from honey

(NOHONCAP) is expected to have a negative and significant relation, since honey income requires less labor input and can, therefore, supplement other household income, but this is not the case. The coefficient is not significant.

Also interesting is that unions and the resources of the family (ADULTMAL) are not significant. These variables were significant in the determination of the number of hectares devoted to sustainable agriculture. The lack of strong estimators suggests that it may not be possible to estimate honey harvest in terms of the number of liters. Instead, it is likely that factors that cannot be measured – such as the availability of flowering plants in the vicinity of the lot – may determine the honey production of the bees.

6.7 SECOND SERIES OF EQUATIONS – EXTENT OF ADOPTION IN HECTARES DEVOTED TO SUSTAINABLE AGRICULTURE

The second series of equations that are estimated are the extent of adoption for APA members. These estimations are made in order to determine how APA farmers respond to the initial adoption of sustainable agriculture. When farmers join APA they are making a commitment to employ sustainable agriculture. A majority of the members produce honey and have intercropping plots, but there are some members who only produce honey. APA members receive free seedlings from the organization but not supplies for honey production. The purpose of these estimations is to determine if there are any specific factors that help to increase the extent of adoption for APA members. These data represent 64 per cent of APA membership and are therefore a representative sample of APA members, but represent a non-random sample of the total population. As a result these coefficient estimates may be biased as well, which needs to be considered when trying to apply these results to the population at large.[17]

The extent of adoption for APA members is measured first by the number of hectares devoted to intercropping and second by the total number of liters of honey harvested, as with the stratified random sample. The extent of adoption is estimated for APA members without first estimating the probability of adopting sustainable agriculture since all APA members harvest honey and/or use intercropping on their lots. Both of the extent equations are estimated using ordinary least squares (OLS) regression.

The first model estimated is the extent of adoption in terms of the number of hectares devoted to intercropping. The R-squared value for this model is 0.41 and the adjusted R-squared value is 0.11, suggesting that only 11 per cent of the model can be explained by the variables included in the regression and that too many parameters may be included in the estimation model. The F statistical value is not significant at the 5 per cent level, suggesting that the overall significance of the model is low (Table 6.3).

The coefficients that are significant in the estimation are the number of unions in which a family participates, not including APA (UNIONAPA), and the number of years that a family lives on a lot (YEARONFM). Both of these coefficients are positive and significant, suggesting that, as the number of unions in which the family participates increases and as the number of years the family spends on the lot increases, so does the number of hectares that are devoted to intercropping. These coefficients suggest that, as a farmer becomes more active in the community and a permanent resident of the lot, the extent of adoption increases. As stated earlier, the seedlings used in the intercropping plots are often distributed through unions. The policies that support sustainable agriculture cannot require that farmers join unions or associations, but the programs could provide the means for these associations to prosper since the influence of local unions can be critical to the adoption decision for many farmers. This raises the importance of the model that estimates the probability of becoming an APA member, which is discussed in the following sections.

It is likely that these results and the estimation of the extent of adoption in terms of honey harvest results are biased, since this group of farmers is not likely to be a random sample of the Ouro Preto do Oeste population, but rather a homogeneous group of families.

6.7.1 Second Series of Equations – Extent of Adoption in Liters of Honey Harvested

The extent of adoption in terms of honey harvest is estimated using an OLS regression. The R-squared value of 0.46 and the adjusted R-squared value is 0.14, suggesting that only 14 per cent of the model can be accounted for by the variables that have been included. The F statistical value is not significant at the 5 per cent level, suggesting that the overall significance of the model is low. Similarly, as in the case of the extent of adoption in terms of the number of hectares, this equation does not provide many significant results (Table 6.3). There are no significant variables. These results may be due to the bias of the sample. This bias is addressed in the following sections.

This model is re-estimated using only three explanatory variables in order to account for the large variance between the R-squared and adjusted R-squared values. The R-squared value is 0.30 and the adjusted R-squared value

Table 6.3 Coefficient Estimates for the Extent of Adoption of APA Members

Variable	Probability of Adoption (Hectares) (n=25) Coefficient	Extent of Adoption (Hectares) (n=25) Coefficient	Extent of Adoption (Honey) (n=25) Coefficient
Constant	-5.3239 (-1.130)	174.31 (0.481)	70.792 (0.367)
OPO	1.1964 (0.881)	-131.95 (-1.060)	
AVEDUHH	0.89694 (0.947)	-10.549 (-0.119)	
AVEDU2	-0.73204E-01 (-0.721)	0.42189 (0.045)	
UNIONAPA	0.89694 * (1.824)	-84.888 (-1.232)	-67.627 (-1.076)
YEARONFM	0.24142 * (2.068)		
YEARSSA	0.26244 (0.494)	69.217 (1.541)	92.384 (2.641)**
PERCAP	-0.15949E-02 (-1.294)		
PERPAS	(-0.542) -0.16944E-01	-2.4507 (-0.888)	
ADULTMAL		10.329 (0.382)	
CATRES		0.89387E-02 (1.114)	
NOHONCAP		0.11520 (0.969)	0.060 (0.672)

Probability of Adoption:		Extent of Adoption (Honey):	
F statistic	1.40	F statistic	2.94
R-squared	0.41	R-squared	0.30
adjusted R-squared	0.11	adjusted R-squared	0.20

Extent of Adoption (Hectares):
F statistic 1.44
R-squared 0.46
adjusted R-squared 0.14

***, **, * indicate significance at the 1%, 5% and 10% levels, respectively.

is 0.20, suggesting that only 20 per cent of the model can be accounted for by the variables that have been included. The F statistic indicates that the model is not significant at the 5 per cent level. The variable that indicates the number of years that the farm family has been using sustainable agriculture (YEARSSA) is significant. The positive sign indicates that, as the number of years increases, the number of liters of honey harvested per year also increases. Although the R-squared and adjusted R-squared values are closer in value, this model is still not a good predictor of the extent of adoption in terms of the honey harvested. As stated earlier, it is likely that the lack of strong estimators may suggest that it may not be possible to estimate honey harvest in terms of the number of liters. Instead, it is likely that factors that cannot be measured, such as the availability of flowering plants in the vicinity of the lot, may determine the honey production of the bees.

6.8 THIRD SERIES OF EQUATIONS – THE PROBABILITY OF ADOPTION

The third series of equations is estimated using the combination of data from the stratified random sample and APA members. These data are combined in order to determine the probability of adopting sustainable agriculture and the extent of adoption for the entire data sample. This additional information is obviously an over-representation of APA members in comparison to the random sample, since 64 per cent of the APA members were interviewed compared to only 5 per cent of the total lots in the settlement. However, the additional information could provide some insight into the adoption of sustainable agriculture. The observations are therefore weighted in order to account for the over-representation. In these regressions the total observations, 196, are weighted according to the ratio of the population and sample probabilities. The weight is calculated as follows: there are 3341 populated lots in Ouro Preto do Oeste. This total can be divided among non-APA (3302) and APA members (39).[18] The total number of interviews, 196, is divided among non-APA members (171) and APA members (25). The weights are calculated as the total population of the group divided by the total interviews of the group. The weight for non-APA members is therefore (39/25)=1.56 and for APA members is (3302/171)=19.31. These weights ensure that the random sample is a representative sample of the total population and that the APA sample is also a representative sample of the total population.

The first two regressions are based on the same estimators of the probability of adopting sustainable agriculture. The second stage of these regressions involves the estimation of the extent of adoption, which is measured in the number of hectares devoted to sustainable agriculture and then in honey harvest. The same variables used to estimate the probability of adopting sustainable agriculture using the stratified random sample are used to estimate the probability of adopting sustainable agriculture using total observations.

The estimation of the probability of adopting sustainable agriculture predicts 90 per cent of the equations correctly. The model correctly predicts 91 per cent of those households that do not adopt sustainable agriculture and also 87 per cent of those households that have adopted sustainable agriculture. The Chi-squared value is 115. At 15 degrees of freedom, the Chi-squared value is significant at the 1 per cent level, indicating that the model has good explanatory power (Table 6.4).

The most significant variable in determining the probability of adopting sustainable agriculture according to this model is the knowledge that sustainable agriculture exists (KNOWSA). Also significant is the union membership (UNION). This coefficient is positive, suggesting that, as the number of unions in which a household participates increases, the likelihood of adoption also increases. The education coefficient is positive and significant at the 10 per cent level, suggesting that, as education increases, the probability of adoption increases. Also significant is the number of years that the family has been on the farm (YEARONFM). The positive coefficient indicates that, as the number of years that the family has been established on the lot increases, the probability of adoption will also increase. This relationship suggests that, as the family establishes permanent property rights on the lot and realizes that the land must sustain the family for future generations, the likelihood of adopting a more sustainable method of farming increases.

As explained earlier, the coefficient estimates provided in Table 6.4 are not comparable with one another because the magnitudes of the coefficients depend on the unit of measurement and because they are expressed as indices. Therefore, the marginal impacts of changes of the independent variables are estimated for the significant independent variables.

The results of these marginal impacts are similar to the probability of adoption that was estimated using the stratified random sample. For example, these marginal impacts show that if education increases by one year, the probability of adopted sustainable agriculture increases by 4.9 per cent, where the probability increase is 3.9 per cent in the random sample estimation (although the variable is not significant in the estimation). Also different in magnitude but similar in sign (compared to the marginal impacts estimated

Table 6.4 Coefficient Estimates for the Probability of Adopting
Sustainable Agriculture and the Extent of Adoption: Total Observations

Variable	Probability of Adopting Sustainable Agriculture (n=196)		Extent of Adoption (Hectares) (n=47)	Extent of Adoption (Honey) (n=47)
	Coefficient	Marginal Impacts[a]	Coefficient	Coefficient
Constant	-6.3926** (-2.431)	-0.838	-0.065464 (0.163)	-835.83 * (-1.747)
NOVA	-0.39772 (-0.808)	-0.047	-1.3078 (-1.639)	185.75 * (1.960)
VALE	-0.35700 (-0.697)	-0.052	-1.2341 (-1.593)	-33.771 (-0.367)
TEIX	0.15275 (0.318)	0.020	-0.25599 (-0.844)	-25.803 (-0.701)
MIR	0.33219 (0.615)	0.044	-0.66287 (-0.833)	159.91 (1.680)
URUPA	-0.49940 (-0.911)	-0.066	-1.7256 (-1.449)	390.58 *** (2.797)
AVEDUHH	0.37189 * (1.933)	0.049	0.10072 (0.239)	59.613 (1.195)
AVEDU2	-0.306-01 (-1.492)	-0.004	-0.204E-01 (-0.413)	-5.9946 (-1.065)
AVEAGEHH	0.11142 (1.177)	0.015	-0.261E-01 (-0.197)	-4.4666 (-0.281)
AVEAGE2	-0.89E-03 (-0.964)	-0.000	0.737E-04 (0.055)	0.16360 (1.015)
UNION	0.39662** (2.365)	0.052	0.56814 (1.654)	37.965 (0.927)
KNOWSA	1.8963*** (4.684)	0.249		
YEARONFM	0.5E-01** (1.973)	0.007	0.13112 ** (2.480)	-2.0438 (-0.326)
YEARSSA			0.17219 (1.179)	30.219 * (1.828)
PERCAP	0.302E-04 (0.474)	0.000	-0.238E-03 (-0.905)	
PERFOR			0.290E-01** (2.195)	2.1400 (1.365)
ADULTMAL			-0.25861 ** (-2.622)	25.420 ** (2.220)

Table 6.4 Continued

Variable	Probability of Adopting Sustainable Agriculture (n=196)			Extent of Adoption (Hectares) (n=47)	Extent of Adoption (Honey) (n=47)
	Coefficient	Marginal Impacts[a]		Coefficient	Coefficient
APA				0.59694 (0.582)	469.56 *** (3.921)
NOHONCAP					0.06004 * (1.980)
LAMBDA				0.29885 (0.411)	110.43 (1.312)

Probability Model:		Extent of Adoption (Honey):	
Likelihood ratio test	115***	F statistic	3.25**
Pseudo R²	0.59	R-squared	0.56
Number of Correct predictions	176 (90%)	adjusted R-squared	0.39

Extent of Adoption (Hectares)[1]:	
F statistic	2.01
R-squared	0.52
adjusted R-squared	0.24

***, **, * indicate significance at the 1%, 5% and 10% levels, respectively.
[a]Marginal impacts are the partial derivatives of $E[y] = \Phi[*]$ with respect to the vector of characteristics. They are computed at the means of the independent variables.

using the stratified random sample) are UNION, KNOWSA and YEARONFM. The variables important to policy implications are knowledge of sustainable agriculture (KNOWSA) and union membership (UNION). The marginal impacts show that when farmers become informed about sustainable agriculture, the probability of adoption increases by 25 per cent, when the family remains on the land for an additional year, the probability of adopting sustainable agriculture will increase by 0.7 per cent, and when a family participates in one additional union, the probability of adoption increases by 5.2 per cent.

6.8.1 Third Series of Equations – Extent of Adoption in Hectares Devoted to Sustainable Agriculture – Based on the Probability of Using Sustainable Agriculture

The extent of adoption for the combined sample is estimated in terms of (1) the number of hectares devoted to sustainable agriculture or intercropping and (2) the number of liters of honey harvested in one year (1996). They are both estimated using the Heckman selection bias correction.

The extent of hectares devoted to intercropping is estimated first. In this model the R-squared value is 0.52 and the adjusted R-squared value is 0.24, suggesting that 24 per cent of the model can be accounted for by the variables that have been included. The F statistical value is not significant at the 5 per cent level, suggesting that the overall significance of the model is low (Table 6.4). This model includes dummy variables for each of the five municipals (as in the first stage). In addition to variables for age, education, union membership, per capita income and the number of years a family has lived on the lot, the number of years a family has practiced sustainable agriculture is included in the estimation. Also, this model includes variables that represent family capital, the number of adult males (ADULTMAL) and the percentage of the lot that is primary forest (PERFOR), indicating how much of the lot may be preserved if sustainable agriculture is adopted.

Neighboring effects do not appear to be relevant to the extent decision, contrary to the same estimation made with the stratified random sample estimation. The inclusion of the APA observations negates the neighboring impacts, suggesting that APA members may be spread throughout Ouro Preto do Oeste, rather than living in the same municipal or clustering in one area.

Also significant are the coefficients of the percentage of the lot that is primary forest (PERFOR), the number of years on the farm (YEARONFM) and the number of males on the farm (ADULTMAL). Those families that do not have a lot of labor on hand (adult males) are choosing to convert more of their land to intercropping compared with those farmers who have greater labor options. The number of adult males living on the lot has a negative and significant coefficient, suggesting that, as the labor available to the family decreases, the likelihood of adopting sustainable agriculture, a less labor-intensive method of farming, also increases.

Education and age of the household heads are not significant in the determination of the hectares devoted to intercropping. It is expected that both of these coefficients would be positive, but in this model they are insignificant.

It is interesting that the variable that indicates the number of years that the family has used sustainable agriculture (YEARSSA) is not significant. This variable was highly significant in the extent of adoption in terms of hectares, which was estimated using the stratified random sample. The inclusion of APA members in the sample has greatly influenced the significance of this variable. This change in significance could be due to the politics that exist in the APA organization. The role that specific members play in the organization may have a greater influence on the number of seedlings that they receive than the number of years of use of sustainable agriculture.

The role that the APA observations play in the model was analyzed by including a dummy variable to represent APA membership. The variable is not significant, suggesting that APA members may not behave differently than other farmers in terms of agricultural choices.[19] All APA members use sustainable agriculture, but if the members do not differ greatly from other farmers it is likely that if non-APA members learn about APA or sustainable agriculture they would adopt sustainable agriculture as well.

6.8.2 Third Series of Equations – Extent of Adoption in Liters of Honey Harvested – Based on the Probability of Using Sustainable Agriculture

The overall significance of the model used to estimate extent of adoption in terms of honey harvest is low. The F statistical value is not significant at the 5 per cent level. The R-squared value of 0.64 and the adjusted R-squared value is 0.43 suggesting that 43 per cent of the model can be accounted for by the variables included. This model is more significant and a better predictor of the extent of honey harvest for APA members than the extent of honey based on non-APA members (Table 6.4).

The most significant variable in the estimation is APA membership. APA membership is expected to be highly significant since all APA members sell and harvest honey. Also significant are the dummy variables for Nova União, Mirante da Serra and Urupá, suggesting that neighboring effects and the spread of information through neighbors is important to honey-harvesting.

The variables showing significance are the municipal dummy variables, the number of years that a family has practiced sustainable agriculture, and APA membership. The coefficients on age of the household heads, education of the household heads and per capita income (from all sources except honey) are not significant, suggesting that honey harvest is influenced greatly by experience. Farming experience can be augmented with APA membership and by increasing the number of years that the family uses sustainable agriculture. This is an important result since experience can be impacted by policy.

6.8.3 Third Series of Equations – Probability of Becoming an APA Member

The probability of becoming an APA member is estimated according to variables that indicate municipal of residence, education of the household heads, age of the household heads, the percentage of the lot that is primary forest, the number of unions in which the family participates and per capita income. This model is used to estimate the value of the Heckman correction used in the extent of adoption equations (in terms of hectares devoted to intercropping and liters of honey harvested). This model can account for any sample selection bias that exists for the APA member observations. The purpose of this estimation is to determine the extent to which the estimations in the second series of equations are biased, if at all.

The Chi-squared value for the model is 134, which is significant at the 1 per cent level for 13 degrees of freedom, suggesting that the model has high explanatory power. The model predicts 88 per cent of the observations correctly. Of these, the model predicts 100 per cent of the non-APA members correctly but only 4 per cent of the APA members correctly. The large difference between predictability of the two groups (non-APA members and APA members) suggests that the predictability of the model is very weak for APA members and raises concern as to whether any of the conclusions from this model are valid, since the model has low prediction ability for the group of observations with which we are most concerned (Table 6.5).

None of the variables used in the estimation are significant at the 1 per cent, 5 per cent or 10 per cent levels. The lack of significant variables suggests that APA members may not be significantly different than non-APA farmers in terms of age, education, per capita income and municipal of residence. The farmers are likely to join APA through family and friend networks rather than being selectively chosen based on characterizations such as income, age or education. APA leaders and founders may have higher levels of education, but this is not true of the remaining members. This result is promising for policy since this suggests that programs that assist in the dissemination of information about sustainable agriculture or provide training in sustainable agriculture are not likely to be adopted by a particular type of farmer (similar to the farmers who joined APA) but rather the average or typical farmer.

6.8.4 Third Series of Equations – Extent of Adoption in Hectares Devoted to Sustainable Agriculture – Based on the Probability of Becoming an APA Member

The extent of adoption in terms of the number of hectares devoted to intercropping by APA members is estimated using variables for education of the household heads, unions in which the family participates besides APA, the number of years the family has resided on the lot, the number of adult

Table 6.5 Coefficient Estimates for the Probability of Becoming an APA Member and the Extent of Adoption

Variable	Probability of Becoming an APA member (n=196)		Extent of Adoption (Hectares) (n=25)	Extent of Adoption (Honey) (n=25)
	Coefficient	Marginal Impacts[a]	Coefficient	Coefficient
Constant	-5.9094 (-0.629)		-4.4655 (-1.066)	271.38 (0.598)
NOVA	0.47007 (0.419)			
VALE	-1.0571 (-0.750)			
TEIX	0.47935 (0.739)			
MIR	0.17700 (0.149)			
URUPA	-1.9450 (-0.956)			
OPO			1.5565 (1.446)	-85.414 (-0.689)
AVEDUHH	0.267E-01 (0.045)		0.73429 (0.990)	26.727 (0.327)
AVEDU2	-0.169E-01 (-0.277)		-0.660E-01 (-0.817)	-0.39793 (-0.044)
AVEAGEHH	0.26264 (0.514)			
AVEAGE2	-0.464E-02 (-0.650)			
UNIONAPA	0.95201 (1.757)		1.1211 (1.303)	-195.59 * (-1.898)
YEARONFM	-0.320E-01 (-0.433)		0.22777 * (2.307)	11.663 (1.091)
YEARSSA			0.33909 (0.924)	76.575 * (2.421)
PERCAP	0.702E-04 (0.410)		-0.173E-02 (-1.666)	
PERFOR			0.228E-01 (1.096)	-0.30789 (-0.126)

Table 6.5 Continued

Variable	Probability of Becoming an APA member (n=196)		Extent of Adoption (Hectares) (n=25)	Extent of Adoption (Honey) (n=25)
	Coefficient	Marginal Impacts[a]	Coefficient	Coefficient
ADULTMAL			-0.23778 (-1.347)	21.412 (1.392)
NOHONCAP				0.793E-01 (0.807)
LAMBDA			-0.22567 (-0.254)	-156.72 (-1.329)

Probability Model:		Extent of Adoption (Honey):	
Likelihood ratio test	134.975***	F statistic	1.47
Pseudo R^2	0.90	R-squared	0.51
Number of Correct predictions	172 (88%)	adjusted R-squared	0.16
Extent of Adoption (Hectares):			
F statistic	1.19		
R-squared	0.46		
adjusted R-squared	0.07		

***, **, * indicate significance at the 1%, 5% and 10% levels, respectively.
[a]Marginal impacts are the partial derivatives of $E[y] = \Phi[*]$ with respect to the vector of characteristics. They are computed at the means of the independent variables.

males on the lot, per capita income, the percentage of the lot that is primary forest and a dummy variable for Ouro Preto municipal.

The overall significance of the model used to estimate extent of adoption, in terms of the number of hectares devoted to intercropping, is low. The F statistical value is not significant at the 5 per cent level (Table 6.5). The R-squared value is 0.46 and the adjusted R-squared value is 0.07, suggesting that only 7 per cent of the model can be accounted for by the variables included. The only variable that is significant in this model is the number of years that a family has resided on the lot. The coefficient of this variable is positive, suggesting that, as the number of years that a family has resided in the lot increases, so does the number of hectares that are devoted to sustainable agriculture. This result is also supported by the previous models discussed in this chapter used to estimate the extent of adoption in hectares (Table 6.5).

The R-squared and adjusted R-squared values differ greatly in this model suggesting that too many explanatory variables have been included in the estimation. The model is therefore estimated using only UNIONAPA, YEARONFM and PERCAP to determine the extent of adoption in terms of the number of hectares that are converted to sustainable agriculture (Table 6.6). The F statistical value is not significant at the 5 per cent level, suggesting the overall significance of the model is low. The R-squared value is 0.31 and the adjusted R-squared value is 0.18, suggesting that only 18 per cent of the model can be accounted for by the variables that have been included. Although the explanatory power of the model is increased, this increase is small. Again, YEARSSA and UNIONAPA are significant.

The estimation of the probability of becoming an APA member does not improve the second series of equations. This model is estimated in order to correct for the censoring bias that occurs when only the APA members are included. The model does not provide any useful information to analysis due to the censoring that may occur.

6.8.5 Third Series of Equations – Extent of Adoption in Liters of Honey Harvested – Based on the Probability of Becoming an APA Member

The extent of adoption in terms of the liters of honey harvested by APA members is estimated using variables for education of the household heads, unions that the family participates in besides APA, number of years the family has resided on the lot, the number of adult males on the lot, per capita income, the percentage of the lot that is primary forest and a dummy variable for Ouro Preto municipal.

The overall significance of the model used to estimate extent of adoption in terms of the number of liters of honey harvested is low. The F statistical value is not significant at the 5 per cent level. The R-squared value is 0.51 and the adjusted R-squared value is 0.16 suggesting that only 16 per cent of the model can be accounted for by the variables that have been included (Table 6.5). The only variables that are significant in this model are the number of unions in which the family participates besides APA (UNIONAPA) and the number of years that the family has been using sustainable agriculture (YEARSSA).

The coefficient of UNIONAPA is negative, suggesting that as the number of unions in which the family participates increases, honey harvest decreases. This is not as would be expected. Since these unions are the main farm organizations in Ouro Preto, they serve as a vital means of information and communication between farmers. The unions have helped to support government agencies in the establishment of sustainable agricultural systems

Sustainable Agriculture in Brazil

Table 6.6 Coefficient Estimates for the Probability of Becoming an APA Member and the Extent of Adoption – Second Estimation

Variable	Probability of Becoming an APA member (n=196)		Extent of Adoption (Hectares) (n=25)	Extent of Adoption (Honey) (n=25)
	Coefficient	Marginal Impacts[a]	Coefficient	Coefficient
Constant	-5.9094 (-0.629)		-2.6211 (-0.984)	503.78 (1.467)
NOVA	0.47007 (0.419)			
VALE	-1.0571 (-0.750)			
TEIX	0.47935 (0.739)			
MIR	0.17700 (0.149)			
URUPA	-1.9450 (-0.956)			
AVEDUHH	0.267E-01 (0.045)			
AVEDU2	-0.169E-01 (-0.277)			
AVEAGEHH	0.26264 (0.514)			
AVEAGE2	-0.464E-02 (-0.650)			
UNIONAPA	0.95201 (1.757)		1.3972 (1.575)	-160.33* (-1.778)
YEARONFM	-0.320E-01 (-0.433)		0.17986* (1.976)	
YEARSSA				80.320* (2.414)
PERCAP	0.702E-04 (0.410)		-0.0006 (-0.698)	
NOHONCAP				0.0754 (0.854)
LAMBDA			0.0842 (0.099)	-136.70 (-1.372)

Table 6.6 Continued

Probability Model:		Extent of Adoption (Honey):	
Likelihood ratio test	134.975***	F statistic	3.19
Pseudo R^2	0.90	R-squared	0.38
Number of Correct predictions	172 (88%)	adjusted R-squared	0.26
Extent of Adoption (Hectares):			
F statistic	2.32		
R-squared	0.31		
adjusted R-squared	0.18		

***, **, * indicate significance at the 1%, 5% and 10% levels, respectively.
[a]Marginal impacts are the partial derivatives of $E[y] = \Phi[*]$ with respect to the vector of characteristics. They are computed at the means of the independent variables.

on farm lots, have disseminated the seedlings used in sustainable agriculture free of charge and have continued to support the use of sustainable agriculture through the sponsorship of annual workshops. It is therefore expected that as participation in these organizations increases, the extent of adoption of sustainable agricultural practices will also increase.

The coefficient on YEARSSA is positive and significant, suggesting that, as the number of years a family uses sustainable agriculture increases, the harvesting of honey will also increase.

The R-squared and adjusted R-squared values differ greatly in this model suggesting that too many explanatory variables have been included in the estimation, as is true for the model of the extent of adoption in terms of hectares of intercropping. The model is therefore estimated using only UNIONAPA, YEARSSA and NONHONCAP to determine the extent of adoption in terms of the number of hectares that are converted to sustainable agriculture (Table 6.6). The model used to estimate extent of adoption in terms of honey harvest is not significant at the 5 per cent level based on the F statistic 3.19 when four explanatory variables are used with 20 degrees of freedom for the numerator. The R-squared value is 0.38 and the adjusted R-squared value is 0.26 suggesting that only 26 per cent of the model can be accounted for by the variables that have been included. Although the explanatory power of the model is increased, this increase is small. Again, the only significant variable is YEARONFM.

The overall results for the estimations involving APA members, including both the probability of becoming an APA member and the extent of adoption, are not significant estimators. The prediction of APA membership is weak and, therefore, not a good means of establishing a bias for the extent equations.

6.9 SUMMARY OF THE ESTIMATION RESULTS AND POLICY IMPLICATIONS

Three series of equations have been estimated in this chapter. The first series of equations estimates the probability of adoption of sustainable agriculture and the extent of adoption once the adoption choice is made using the stratified random sample. The first series of equations clearly performs well. Both the adoption and extent of adoption have high explanatory power and many significant variables. The third series of equations makes the same estimation with a combination of data collected from APA (a farmer organization that supports the use of sustainable agriculture) and the random sample. The third series of equations also has high explanatory power and many significant variables. The second series of equations as well as the estimation of the probability of becoming an APA member do not perform as well. The reason that these equations do not perform well is likely to be the result of the inclusion of the APA member data. Data gathered from this homogenous group of farmers do not lead to significant regression estimates.

The conclusions from the models involving APA members are: APA members do not significantly differ from non-APA farmers in the characterizations examined in this study. There are no significant indicators that predict which farmers are more likely to become APA members. Instead, it is concluded that the probability of becoming an APA member is determined by variables that cannot be measured, such as the politics involved in the recruiting of members. It is likely that the ideology that exists in the organization, which cannot be measured in a regression, plays a large role in the extent of adoption.

The impacts that APA has made on adoption and the extent of adoption could be measured best by a larger data sample that includes similar farmer organizations in different settlements of the Amazon. These data do not include observations from areas outside of Ouro Preto do Oeste because of budget and time constraints. A larger data sample would provide the ability to compare the impact of farmer organizations of farm choice in a variety of situations.

Although the different models that have been estimated in this chapter have produced different conclusions and have used different data groups, there are some similarities between the results, and these results are important to mention. The probability of adopting sustainable agriculture is highly predictable in the first and third series of estimations and is influenced greatly by the knowledge that sustainable agriculture exists and by the number of unions in which a family participates. This result is important to policy implications since both of these factors may be influenced by economic and social programs.

An important variable in the estimation of the extent of adoption is the number of years that the family has practiced sustainable agriculture, indicating that once sustainable agriculture is adopted, farmers are likely to increase the extent to which it is practiced throughout the years. Recognizing this variable is also important to policies because this continued practice of sustainable agriculture by families implies that policies that provide initial assistance and seedlings can lead to multiple effects for each farmer.

Another conclusion that can be drawn from these models is that as families become permanent residents of their lots, they become more likely to adopt more sustainable means of production. Property rights have become secure over the past ten years in Ouro Preto do Oeste, and these rights have instilled an incentive to maintain farm lots and the productivity of these lots for future generations.

The marginal impacts are calculated for both of the equations that estimate the probability of adoption. The marginal impacts for these models in terms of policy implications are similar. The variables important to policy implications are the knowledge of sustainable agriculture and union membership. The impact that the knowledge of sustainable agriculture makes on the probability of adoption is very large, and implies that policies that can help increase this knowledge can impact the adoption greatly. Programs likely to be most successful are those that concentrate on the dissemination of information about new agricultural methods and provide training for the farmers in the methods of sustainable agriculture.

ENDNOTES

1. It is difficult to compare the labor required for each dollar of income derived from intercropping and sustainable agriculture, since income varies greatly according to the crop harvested. For example, one hectare of rice yielded an average $251, while one hectare of beans averaged $264, and manioc averaged $2507 per hectare. Comparing this to intercropping plots is also difficult because many of the crops produced in the intercropping areas are perennials that are native to the area, but not yet sold in developed markets (as rice, beans and manioc are). However, one can compare the average labor used per year for each. This can be an appropriate way to compare the two methods of agriculture, because it was determined that there was no significant difference between the income before and after using sustainable agriculture for those farmers who had once used slash-and-burn agriculture and switched to sustainable agriculture. But, it was found that there was a significant difference in the amount of labor that was required before and after the switch. Therefore, it was concluded that these farmers may be trading labor for leisure by adopting the new agricultural technique. In a sample of 25 farmers, it was determined that the burning of land to plant annual crops – in combination with planting, maintaining and harvesting annual crops –

required 55 per cent more hours of work per hectare than the planting, harvesting and maintaining of one hectare of intercropped annuals and perennials. Complicating the comparison is the labor requirement for the harvesting of honey. Honey harvest is included in the definition of sustainable agriculture since this income source is not dependent upon the destruction of the forest. This crop is difficult to compare to the yield from slash-and-burn agriculture because it cannot be measured in hectares. Since, of those farmers who sell honey, it represents a greater portion of income and requires less labor per year, it can be concluded that the farmer who adopts this income source can devote less time per year to the honey harvest than to slash-and-burn agriculture. In a sample of 25 farmers, it was determined that the labor required to harvest the average number of liters of honey (harvested in one year) was 74 per cent lower (in days per year) than the labor required to burn, plant, harvest and maintain one hectare of annual crops (below the average of 2.5 hectares harvested per year), and 41 per cent lower than labor required to plant, harvest and maintain one hectare of intercropped annuals and perennials. On average, the income from honey represents 18 per cent of the total income of those farmers who produce honey (APA members), while annual crops represent 10 per cent of total income.

2. A majority of the lots in Ouro Preto do Oeste are 100 hectares in size.
3. Sustainable agriculture includes the intercropping of perennials with varying perennials and/or annuals, alone or in combination with income sources such as beekeeping and aquaculture, which do not rely on the destruction of the rain forest. A farmer is counted as adopting sustainable agriculture if the family has intercropping plots and/or harvests honey. Fish-raising is not included in the analysis since only one farmer cultured fish during the interview year (1996). Eleven additional farmers stated that they began to raise fish in 1996, but would not harvest in that year.
4. The choice to maintain annual crops in addition to adopting sustainable agriculture reduces the risk associated with the new technology. It is assumed that once the benefits of sustainable agriculture are realized, the farmers will not need or desire to maintain annual crops.
5. Where $u_i = (e_{1i} - e_{2i})$ as is derived in equation 6.3.
6. See Section C.5, 'The Tobit and Heckman Models Compared', in Appendix D for a detailed comparison of the Heckman and tobit models.
7. See Section C.4, 'Heckman Model Specification', in Appendix D for the mathematical framework of the Heckman model.
8. The Heckman model (Heckman 1979) allows for different variables that determine the adoption choice and then the effort allocated when adoption is chosen. The two-equation procedure involves the estimation of a probit model of the adoption decision, calculation of the sample selection control function, and incorporation of that control function (the inverse Mills ratio) into the model of effort, which is estimated with ordinary least squares (OLS).
9. Most studies estimate the extent of adoption using the tobit model since the elasticities evaluated at the means can be decomposed into the elasticity of adoption and the elasticity of effort given that adoption occurs (Adesina and Zinnah 1993, Gould *et al.* 1989, Norris and Batie 1987). These decomposed elasticities may be interesting for the interpretation of policy since they provide information about the elasticity of adoption and the elasticity of effort, but it is asserted here that the assumption that is necessary to calculate these elasticities is not relevant to this research and therefore the elasticities would not provide any additional information.
10. See Section C.3.1, 'Tobit Model Elasticities', for the derivation of the decomposition of the elasticities.
11. Since sustainable agriculture is defined as the use of the intercropping of perennials with varying perennials and/or annuals, alone or in combination with income sources such as beekeeping and aquaculture, which do not rely on the destruction of the rain forest, the extent

of deforestation is measured in these terms. Fish-raising is not measured in the extent equations, since only one farmer cultured fish during the interview year (1996). Eleven additional farmers stated that they began to raise fish in 1996, but would not harvest in that year.

12. This sample is stratified in order to ensure that each of the six municipals which lie in Ouro Preto do Oeste are represented equally based on the number of farm lots in each municipal. The percentage of the total interviews from each municipal represents the percentage of lots that lie in the municipal. After the number of interviews to be conducted in each municipal was determined, a random sample was conducted in each municipal.
13. One observation is dropped in the regression analysis since it was found that the observation was an outlier that dominated the estimation results.
14. The municipal Ouro Preto is not included as a dummy to avoid collinearity between the variables and to test for any significant difference between the other five municipals and Ouro Preto. Ouro Preto was chosen as the reference municipal because it is the municipal located in the center of the settlement and is assumed to be average in terms of agriculture, income and lot size, but different in terms of population and development. Since this municipal was settled first, it is the oldest municipal with the highest urban population. The urban population is not used in the analysis.
15. The household heads are the main decision-makers on the farm lots.
16. There have been many attempts to derive goodness of fit measures for discrete choice models (Greene 1997). An analogue to the R-squared used in conventional OLS regression models has been developed and is the most popular of the fit measures for binary choice models. The pseudo-R-squared or the likelihood ratio index (LRI) was developed as a fit measure for probit, logit and tobit models (Gujarati 1995). This measure has an intuitive appeal in that it is bounded by 0 and 1. There is no way for the LRI to equal 1, but it can come close (Greene 1997). Values between 0 and 1, however, have no natural interpretation (Greene 1997). This value will be presented in the following models, since it is commonly referred to, but it will not be discussed further since it is lacking natural interpretation. The LRI index for this model is 0.33.
17. This bias is addressed in the third series of equations where first the probability of becoming an APA member is estimated and then the extent of adoption using a Heckman correction to adjust for the censored sample.
18. At the time of the interviewing there were 39 active members of APA; however, this number has increased since 1996.
19. The differences between APA members and non-APA members is also addressed in the estimation of the probability of becoming an APA member (Section 6.8.3).

Chapter 7

Conclusion

7.1 MODEL AND METHODS

Tropical deforestation in Brazil is the result of market failures occurring on global, national and local levels. Market failure, on all levels, impacts the local farmers who live in the tropical forests and depend on the land for their livelihoods. This study investigates the decisions farmers make in relation to farming technique alternatives and how these decisions impact deforestation.

The adoption of sustainable agriculture is estimated in a technology framework where the choice of using a new farming technique, sustainable agriculture, or the older technique, slash-and-burn agriculture, is treated as a dichotomous choice. A two-stage Heckman model is estimated to determine, first, the probability of adopting sustainable agriculture and then the intensity of adoption once sustainable agriculture is used. The results from the analysis are used to make policy suggestions as to the most effective ways to increase the adoption rate of sustainable agriculture. In particular, the roles of farm unions and information dissemination are investigated since these factors may clearly be influenced by policy.

7.2 RESULTS

One important result of the analysis is that information plays a vital role in the adoption of sustainable agriculture by assisting local farmers in the adoption of a superior method of farming. This result is consistent with the results of Hussain *et al.* (1994), D'Souza *et al.* (1993), Feder and Slade (1984), and Opare (1977). Also found to be important is the role of human capital. As with Strauss *et al.* (1991), Lin (1991), and Akinola and Young (1985) access to education is found to increase the rate of adoption. The farmers who have a greater number of completed years of education are found to be more likely to adopt sustainable agriculture, although this result is not supported by all of the estimations.

This study adds to this body of literature by investigating the role that farm unions and local farm organizations play in the adoption of a new technology. It is found in the regression analysis that unions and local farm organizations play a critical role in the adoption of sustainable agriculture. This result is important since these organizations can be influenced by policy. For example, financial and educational support of sustainable agriculture could be provided through these organizations.

In addition, this analysis adds to the body of literature because it is a unique look at the influence of capital assets. Previous studies have suggested that as capital assets increase the adoption of new technologies will also increase (Miller and Tolley 1989, Holden 1993, Kebede *et al.* 1990, Jaeger and Matlon 1990, Norris and Batie 1987, Savagado *et al.* 1994, Rahm and Huffman 1984, Shakya and Flinn 1985), but this analysis finds that the extent of adoption is influenced greatly by a lack of capital assets such as available labor (measured by the number of adult males). The families that have the fewest capital assets are more likely to adopt more intensely the agricultural technique that requires the least labor, suggesting that more farmers may adopt sustainable agriculture if the smaller labor requirements (relative to slash-and-burn) are learned.

One main issue discussed is that sustainable agriculture is not being adopted by many farmers in Ouro Preto do Oeste despite potential increases in income and welfare. Only 1 per cent of the farmers who live in Ouro Preto do Oeste have adopted sustainable agriculture since it was first introduced to the city in the early 1980s through governmental programs.

It is concluded that the main reason for the low adoption rate of sustainable agriculture in Ouro Preto do Oeste is imperfect information. A majority of farmers interviewed did not have any knowledge of sustainable agriculture. Only 21 per cent of the total farmers interviewed knew about or had training in the use of sustainable agriculture. Of the farmers who had knowledge of sustainable agriculture (including APA members) 83 per cent chose to adopt sustainable agriculture. In the stratified random sample of 171 farmers (not including APA members), only 10 per cent of the farmers knew about sustainable agriculture and of those farmers who knew about it, 60 per cent of them adopted it. These results suggest that the adoption of sustainable agriculture could be increased considerably through programs that support the dissemination of information about sustainable agriculture.

Another important result is that there are few significant variables in the estimation of the probability of becoming an APA member. The lack of significant variables suggests that APA members may not be significantly different than non-APA farmers in terms of age, education, per capita income and municipal of residence. The farmers are likely to join APA through family and friend networks rather than being selectively chosen on the basis of characterizations such as income, age or education. APA leaders and founders may have higher levels of education, but this is not true of the remaining members. This result is promising for policy since it suggests that programs that assist in the dissemination of information about or provide training in sustainable agriculture are not likely to be adopted by a particular type of farmer (similar to the farmers who joined APA) but rather by the average or typical farmer.

The overall results of this research are promising. Many farm families have established property rights and have recognized that it is necessary to maintain the nutrient rich soils for future generations. The lots are generally passed on to future generations that depend on productivity of the land in order to survive, and this is a reason why many of the farmers have adopted sustainable agriculture. As the land that can support agriculture is depleted through burning, families have begun to realize the importance of sustainable agriculture. It is therefore important to educate more families about alternative land use methods. Programs supporting the dissemination of information are therefore likely to be successful in encouraging more families to adopt more sustainable methods of farming.

7.3 FUTURE RESEARCH

Further research of this area is necessary in order to make inferences about the success of sustainable agriculture. Since the adoption of this agricultural method is relatively new to the region, there are not many conclusions to be drawn as to the success of sustainable agriculture and whether intercropping can maintain the soil quality necessary to support agricultural crops for extended periods of time. It will therefore be interesting to continue this study and determine the success of these agricultural systems in terms of income and crop yields as Ouro Preto do Oeste becomes more developed, and to investigate the impact that the development makes on deforestation and the decisions by farmers to adopt sustainable agriculture.

Appendix A

Operation Amazonia

A.1 OPERATION AMAZONIA

The Brazilian government began colonization programs in order to establish settlements in the undeveloped Amazon region in response to the many social and political problems in the populated areas of the northeast and southeast. A major component of the government programs was road development and improvement. Road construction began in the states of Goiás, Maranhão, Mato Grosso and Pará. Responding to infrastructure development and settlement in the tropical forests of Peru, Venezuela and other adjacent countries, the military government, which came to power in 1964, launched 'Operation Amazonia' (Southgate 1994, Southgate 1992, Dale and Pedlowski 1992, Hecht and Cockburn 1990, Guppy 1984). Among the first infrastructure projects was the first 1200 km of the Tranzamazon Highway, which runs west from the northeastern state of Maranhão and continues across the Amazon region to western Amazonas. It was completed in 1972 (Southgate 1992).

Agriculture colonization was encouraged by the government along the new highways. The primary objective of the completion of the Tranzamazon Highway was to facilitate the resettlement of 70 000 families from poverty in the northeast region. The area was suffering economically and socially from a long drought (Frohn et al. 1990, Moran 1990). The colonization objectives were not met because of the low fertility of the soil along the highway. Only about 3 per cent of the land along this highway is suitable for agriculture, resulting in the settlement of only 8000 families by 1980, which was well below the government's objectives (Browder 1994, Southgate 1992, Hecht and Cockburn 1990).

By contrast, in the areas where conditions were hospitable to agriculture, road construction induced rapid development of agricultural settlements, such as the completion of BR-364 in 1984 in Rondônia and BR-163 in Pará and Mato Grosso. In these areas the deforestation rate became the highest in Brazil only after the completion of the highways (Figure 2.2).

Inappropriate government planning resulted in economic and environmental disasters in the area surrounding the Tranzamazon highway, BR-163 and BR-364. The plans were implemented quickly and recklessly, resulting in a great deal of destruction both to the tropical forests and to the welfare of many people involved. For example, government planners and extension workers based the agricultural and economic development of the settlement areas on annual crops (particularly upland rice), which are generally considered to be environmentally and ecologically unsustainable in areas cleared from tropical rain forests, except under very favorable conditions (Browder 1994, Fearnside 1983, Goodland 1980).

Under the National Integration Program (PIN), which fostered the construction of the Tranzamazon Highway and the settlement of the area, the Tranzamazon settlers found themselves in circumstances that were far from favorable. First, the colonization projects were great distances from major markets for agricultural commodities, which put the settlers at a disadvantage compared with other more efficient producers elsewhere in Brazil. Second, largely because of the high cost of transportation due to poor road conditions, fertilizers, pesticides and herbicides were sold in the region at prices beyond the reach of the farmers. Without these modern inputs, crops often succumbed to pests and diseases (Hecht and Cockburn 1990, Mahar 1989).

The government integrated the construction of BR-364 in Rondônia with a World Bank-supported settlement program (POLONOROESTE) designed to relocate small-scale farmers from other regions of Brazil (Browder 1988). The combination of poor land reform programs in other parts of the country, relatively fertile soils and an enormously effective government campaign resulted in an explosive growth in the region, unanticipated by the government (Anderson 1993, Mahar 1989). Rondônia's rapid population growth and uncontrolled settlement has had devastating effects on the rain forest. The speed at which this deforestation took place was, in some areas of the state, truly astonishing. Most of the clearing was done for agricultural purposes (Mahar 1989). The government's provision of subsidies to farmers and ranchers continued up until the 1980s (Dale and Pedlowski 1992).

A.2 MIGRATION TO THE AMAZON

The combination of government programs, diverging land prices and the poverty in urban areas of the country resulted in a massive migration movement in the 1970s and 1980s into the Amazon Basin of Brazil. Deforestation has increased as roads have made the region more accessible to settlers, farmers, loggers, miners and ranchers.

In 1960 much of the Brazilian Amazon could only be reached by small aircraft or river boats. The tropical forests were largely undisturbed. In the late 1960s a large number of colonists from the south, southwest and northeast began to move to the Amazon Basin responding to government incentives. Many settlers were attracted to the frontier by the prospect of owning land; lots up to 100 hectares were supplied by the government (Baer 1995). The new settlers then cleared the forest for their settlements, farms and ranches (Dale *et al.* 1993, Pedlowski and Dale 1992, Hecht and Cockburn 1990).

Human migration to the Amazon greatly increased the deforestation rate, but unlike tropical forests in other parts of the world, population pressure, by itself, has probably not been a major factor. Instead incentives provided by the government through Operation Amazonia have led to most of the deforestation in the area. In Brazil most rural poor have been much more likely to emigrate to the large cities on the eastern coast of the country rather than to the Amazon. In the 1970s at the start of Operation Amazonia, 16 million rural inhabitants migrated to cities compared to only 770 000 to the Amazon frontier (Mahar and Schneider 1994) resulting in high rates of urbanization.

A.3 SUBSIDIES

Road construction is the most visible aspect of Operation Amazonia, but the creation of the Superintendency for the Development of Amazonia (SUDAM), established to provide subsidies for agricultural development, also indirectly added to deforestation in the area. The Bank of Amazonia (BASA) was founded to provide and distribute the funds for Operation Amazonia (Southgate 1992). In 1963 legislation was passed which induced Brazilian firms to invest in the Amazon Basin by reducing their tax payments by half if they invested the money in that region.

The Brazilian government generally promoted cattle-ranching with the subsidy program. The subsidies administered to the settlers were often as great as 75 per cent of the cost of starting a ranch (Holloway 1993). By late 1985 SUDAM had approved 950 projects in the Amazon, 631 of which were cattle ranches mostly located in the eastern Amazonian states of Maranhão and Pará. Between 1974 and 1980 the volume of subsidized rural credit committed to Amazonia increased almost tenfold in real terms. Many of these projects would not have been profitable in the absence of subsidies (Mahar 1989). On the other hand, the impact of subsidies should not be exaggerated, since Mahar (1989) found that these projects were responsible for no more than 10 per cent of the total land clearing.

A.4 DIVERGENCE OF LAND VALUES

Land prices in the southern regions of Brazil began to increase rapidly relative to the north in the 1970s. In 1970 a farmer (or anyone) in the south could purchase about twice the amount of land in the north as in the south for the same price (Southgate 1992). In 1975 one hectare of land in the southeast could be traded for 13 hectares in the Amazon. Within ten years land in the south was worth up to ten times that in the north (Southgate 1992). Although increased production in the south can explain some of the increase in land value, this increase combined with government incentives – which provided many lots in the settlement programs free of charge (through Operation Amazonia) – resulted in many farmers moving north.

A.5 LAND TAX SYSTEM

The rural land tax in Brazil, administered by the national government, is currently assessed at a maximum of 3.5 per cent of the market value of the land, and 50 per cent of the total land owned is exempt from taxation (Mahar and Schneider 1994). Reductions of up to 90 per cent of the basic rate are given according to the degree of utilization of land (the proportion cleared and efficiency indicators such as crop yields and cattle stocking rates). This tax system provides the incentive to clear land for pasture and agriculture. Fortunately the tax structure has had little influence on the rate of deforestation mainly because the landowners declare the value and efficiency of its use themselves. For example, in 1996 only half of the farmers in Rondônia paid any land tax and, of the 50 per cent who did, the average payment was only equal to $5 (Mahar and Schneider 1994), which is rarely paid anyway (Moran 1993). In 1997 it was estimated by the labor union of rural workers (STR) that only approximately 40 per cent of farmers pay their land taxes in Ouro Preto do Oeste.

A.6 AGRICULTURAL INCOME TAX

The agricultural income tax system supplies another incentive to deforest. Corporations and individuals can exclude up to 80 per cent and 90 per cent, respectively, of agricultural profits from their taxable income. Corporate agricultural profits are taxed at a rate of only 6 per cent. Combined with the depreciation provision the tax can be as low as 1.2 per cent. Comparatively, corporate income taxes from other sources are subject to a rate of 35-45 per cent (Binswanger 1991). The incentives of this tax structure are to entice corporations to invest in agriculture.

This tax structure may seem irrelevant to the small farmer who does not have to pay income taxes, but since poor individuals cannot benefit from the tax breaks as corporations can, the tax benefits affect small farmers indirectly. Since agricultural income is taxed at a lower rate than other forms of income, agriculture is treated as a tax shelter, which leads the market price of land to contain a component capitalizing on these profits, which increases the relative price of land (Binswanger 1991). The corporations therefore drive up the price of land as they invest in this relatively cheaper capital asset, making land more expensive for the small farmer.

Appendix B

English Survey

Date: _____

I. General Information

1.1 County_____   1.2 Area _____

1.3a Road _____   1.3b Lot Number _____

1.4a Area _____
    1.4b distance to city center _____
    1.4c distance of paved road _____
    1.4d distance of unpaved road _____

1.5 Number of lots _____ 1.6 Population _____ 1.7 APA members _____

1.8 How old are you? _____

1.9a Did you go to school? Yes No

1.9b How many years did you study? _____

1.10 How many years have you been a farmer?_____

1.11 How many years have you lived on this lot? _____

1.12 Are you the owner of the lot? Yes No

1.13a What is the size of your lot (in hectares)? _____

1.13b How much of your lot is being used for:
    1. Agriculture _____
    2. Pasture _____
    3. Forest _____
    4. Sustainable Agriculture _____

1.14a How did you acquire this lot?   1.14b How are the roads to your
    a. Occupied the land   lot during the rainy season?
    b. Land was given to you through INCRA   1. Good
    c. Paid with a loan from the bank   2. Passable
    d. Bought with your own money   3. Impassable
    e. Occupied the land and then bought it

1.15a Did you participate in any unions or associations for farmers?
        Yes No
1.15b How many? _____ which? _____

1.16 What state were you born in?_____

1.17 What state were your parents born in? _____

1.18 Were (Are) your parents farmers? Yes No

II. Family Characteristics

2.1 Who lives on this farm? How many people? _____

	number
a. men > 10	
b. men < 10	
c. women > 10	
d. women < 10	

How old is the woman of the house? _____
For how many years did she go to school? _____

2.2 How many of the following items does your family own?

    a. Horses_____   b. Chickens_____

    c. Cattle _____   d. Pigs _____

    e. Televisions_____   f. Sheep_____

    g. Satellite Dish _____   h. Goats _____

    i. Tractors _____

    j. Equipment for use with animals _____

    k. Cars_____   l. Trucks _____

    m. Motorcycles _____   n. Wheel Barrel _____

    o. Telephones _____   p. Refrigerators _____

    q. Bicycles _____   r. Other lots _____

    s. Other houses_____   t. Workers _____

    u. Chain Saws _____   v. Garden (size) _____

    w. Other types of equipment _____

2.3 Why did your family migrate to Rondônia?
    a. The land was free
    b. For work
    c. Because other members of your family migrated to Rondônia
    d. Land
    e. Other (specify)

2.4 What year did your family migrate to Rondônia? _____

III. Methods of Cultivation

3.1a How often do you slash-and-burn your lot to plant crops?
    a. One time
    b. Every two years or more
    c. Every year
    d. Never

3.1b What land?
    a. Primary forest
    b. Secondary forest

3.2a How often do you slash-and-burn your lot to create a fire line?
    a. One time
    b. Every two years or more
    c. Every year
    d. Never

3.2b What land?
    a. Primary forest
    b. Secondary forest

3.3a How often do you slash-and-burn your lot to create pasture?
    a. One time
    b. Every two years or more
    c. Every year
    d. Never

3.3b What land?
    a. Primary forest
    b. Secondary forest

3.4a How much of your land did you burn this year? _____

3.4b Last year?_____

3.5 What type of equipment do you use to clear and burn your land?
_____

3.6 How many people help to clear and burn your lot? _____

3.7 How many days do you need to clear and burn your lot? _____

IV. Production

4.1 The following questions are about your harvest of annual crops
    a. How much _____ did you plant this year (in hectares)?
    b. Last year? _____
    c. How much _____ did you harvest last year?
    d. How much _____ did you sell last year?
    e. At what price? _____

Annuals	1. Area 1996	2. Area 1995	3. Unit	4. Harvest 1995	5. Unit	6. Sell Y/N	7. How many	8. Price
a. Rice								
b. Corn								
c. Beans								
d. Manioc								

4.2 a. Do you use chemical fertilizers? Yes No For which crops?
    b. Do you use organic fertilizers? Yes No For which crops?
    c. Do you use pesticides? Yes No For which crops?
    d. Do you use herbicides? Yes No For which crops?

	Rice	Corn	Beans	Manioc	Coffee	Garden	
Chemical Fertilizers							
Organic Fertilizers							
Pesticides							
Herbicides							

4.3 The following questions are about your harvest of fruits and trees.
   a. How much _____ did you plant this year?
   b. Last year? _____
   c. How much _____ did you harvest this year?
   d. How much _____ did you sell this year?
   e. At what price? _____

Perennial Crops	1. Area 1996	2. Area 1995	3. Unit	4. Harvest 1995	5. Unit	6. Sell Y/N	7. How many	8. Price
a. Coffee								
b. Cacao								
c. Banana								
d. Citrus								
e. Cupuaçú								
f. Pupunha								
g. Açaí								
h. Acerola								
i. Mahogany								
j. Valuable wood								
k. Other Wood								
l. Jake fruit								
m. Abacati								
n. Mango								
o. Coconut								
p. other								

4.4 Where do you sell your crops? _____

V. Cattle and Pasture

5.1 About your cattle, how many of these produce milk? _____

5.2 Does anyone pick up your milk to sell at a market?

5.3 What is the importance of milk to your income?
        1. Very important
        2. Important
        3. Not very important
        4. Not important

5.4 How much milk do you produce each day?
        5.4a in the dry season _____
        5.4b in the rainy season _____

5.5 How much do you sell your milk for? _____

VI. Sustainable Techniques of Agriculture

6.1 Do you know about sustainable techniques of agriculture? Yes No

'Sustainable agriculture contributes to the production of agricultural products at the same time as it conserves the tropical rain forest. Sustainable agricultural methods include the use of various perennials planted together and the use of perennials with annual crops such as corn, rice, beans and coffee. This agricultural method uses the forest while reducing the need to burn. In addition, the farmers who use these methods do not depend totally on the income raised from cattle (in the form of meat or milk), but instead have various other forms of income like beekeeping and fish-raising. Therefore the majority of income of these farmers is derived from perennial crops, beekeeping and fish-raising, all of which are sustainable uses of the forest and are dependent upon the standing forest'.

6.2 Do you use any of these techniques? Yes No If yes, which? If not, go to question 6.8

6.3 a. Planting perennials
     b. Leaving trees standing
     c. Intercropping (only perennials) Which trees?
     d. Intercropping (perennials and annuals) Which trees and annuals?
     e. Using other forms of income

6.4 When did you begin using these new techniques?_____

6.5 How much of your lot do you use for the following?
     a. Planting perennials _____
     b. Leaving standing trees _____
     c. Intercropping (perennials) _____
     d. Intercropping (perennial and annuals) _____
     e. Using other forms of income _____

6.6 What other forms of income do you have?
     a. Beekeeping
     b. Fish-raising
     c. Other (specify)

6.61a How much honey did you harvest last year? _____
6.61b Sell? _____
6.61c At what price did you sell? _____

6.62a How much fish did you harvest last year? _____
6.62b Sell? _____
6.62c At what price did you sell? _____

6.7 What is the importance of _____ for your income?
     1. Very important
     2. Important
     3. Not very important
     4. Not important
a. Beekeeping _____
b. Fish-raising _____
c. Other (specify) _____

6.8 Do other people who live on this farm work outside or have other forms of income?

6.8a How many people? _____

6.8b What do they do?_____

6.9 Why do you use these new techniques of farming?
    a. It is more profitable
    b. Seed was provided
    c. Money was provided
    d. It's better for future generations
    e. It is better for the forest and the region
    f. My parents use these methods
    g. My neighbors use these methods
    h. Other (specify)

VII. The Cost of Sustainable Agriculture (for those farmers who use it)

7.1a Do these new techniques require more or less work? Yes No

7.1b If yes, how much more or less?

7.2a Are these new techniques more expensive? Yes No

7.2b If yes, how much more?

7.3 Do you use the same area of your lot for these new techniques? Yes No

7.4 How much did you harvest before you began using these new
techniques (6.4) _____ years ago?

Crop	Old Harvest	Unit
a. Rice		
b. Corn		
c. Beans		
d. Manioc		
e. Coffee		
f. Cacao		
g. Banana		
h. Citrus		
i. Cupuaçú		
j. Pupunha		
k. Açaí		
l. Acerola		
m. Mango		
n. Coconut		
o. Jake fruit		
p. Abacati		
q. Passion fruit		
r. other		
s. other		
t. other		
u. other		

VIII. Methods of Agriculture

8.1 Which of the following items influence you, and in what degree, as to your choice of agricultural methods on this farm?

1. A lot of influence
2. Some influence
3. No influence

a. Your neighbors _____
b. An association, union, or cooperative _____
c. Free Seed _____
d. Free trees _____
e. Bank Loans _____
f. A market to sell your annual crops _____
g. A market to sell your perennial crops _____
h. The soil and soil quality _____
i. Precious Wood _____
j. The cultivation methods of people in Ouro Preto do Oeste _____
k. The cultivation methods of your parents _____

Surveyor's Comments:

Appendix C

Portuguese Survey

Número _____

Data: _____

I. Informação Geral

1.1 Cidade _____   1.2 Gleba _____

1.3a Linha _____   1.3b Lote _____

1.4a Área _____
    1.4b distância _____
    1.4c via pavimentada_____
    1.4d via sem pavimento _____

1.5 Número de lotes _____ 1.6 População _____ 1.7 Número na APA _____

1.8 Quantos anos o Sr. tem? _____

1.9a O senhor estudou? Sim Não

1.9b Quantos anos o senhor estudou? _____ Até que série estudou? _____

1.10 Há quantos anos o senhor trabalha como agricultor?_____

1.11 Há quantos anos o senhor trabalha nesta propriedade? _____

1.12 O senhor é o proprietário do lote? Sim Não

1.13a Qual o tamanho do seu lote (em hectares)? _____

1.13b Quanto de seu lote está sendo usado para
      1. Agricultura _____
      2. Pasto _____
      3. Floresta _____
      4. Outro _____

1.14a Como adquiriu este lote?   1.14b Como são as estradas para o
    a. Ocupou a terra   seu lote na época de chuva?
    b. Terra foi doada pelo INCRA   1. Bom
    c. Pegou financiamento do banco   2. Mais ou menos
    d. Comprou com o seu próprio dinheiro 3. Ruim
    e. Ocupou a terra e depois comprou

1.15a O senhor pertence a alguma associação, sindicato ou cooperativa?
    Sim Não
1.15b Qunanto? _____ Quiais? _____

1.16 O senhor nasceu em que estado?_____

1.17 Seus pais nasceram em que estado? _____

1.18 Seus pais eram agricultores? Sim Não

II. Características da Família

2.1 Quem mora nesta (na sua) propriedade rural? Quantas pessoas? _____

	numero
a. homens > 10	
b. homens < 10	
c. mulheres > 10	
d. mulheres < 10	

Quantos anos a senhora tem? _____

A senhora estudou por quantos anos? _____

2.2 Que bens a sua família possui?

a. Cavalos_____   b. Frangos_____

c. Gados _____   d. Porcos _____

e. Televisões_____   f. Carneiros_____

g. Antenas parabólicas ____   h. cabras _____

i. Tratores _____

j. Equipamento para tração animal _____

k. Carros_____   l. Caminhões _____

m. motocicletas _____   n. Carrinhos-de-mão _____

o. Telefones _____   p. Geladeiras _____

q. Bicicletas _____   r. Outros lotes _____

s. outras casas_____   t. trabalhadores _____

u. motoserras _____   v. horta (tamanho) _____

w. Outros tipos de equipamento _____

2.3 Por que sua família veio para Rondônia?
   a. O lote foi dado de graça.
   b. Para trabalhar
   c. Porque outros membros de sua família vieram para Rondônia
   d. Terra
   e. Outro (qual)

2.4 Quando a sua família chegou em Rondônia? _____

II Métodos de Lavoura

3.1a Com qual freqüência você roça e queima para plantar?
    a. Um vez
    b. A cada dois anos ou mais
    c. A cada ano
    d. Nunca

3.1b Qual terra?
    a. floresta primeira (mata)
    b. floresta secundária (capoeira)

3.2a Com qual freqüência você faz acero?
    a. Um vez
    b. A cada dois anos ou mais
    c. Cada ano
    d. Nunca

3.2b Qual terra?
    a. floresta primeira (mata)
    b. floresta secundária (capoeira)

3.3a Com qual freqüência você roça e queima para plantar pasto?
    a. Um vez
    b. A cada dois anos ou mais
    c. Cada ano
    d. Nunca

3.3b Qual terra?
    a. floresta primeira (mata)
    b. floresta secundária (capoeira)

3.4a Quanto da área do seu lote você queimou este ano? _____

3.4b E no ano passado?_____

3.5 Que equipamento você usa para roça e queimar o seu lote?

_____

3.6 Quantas pessoas fazem a preparação para as queimadas dentro o lote?__

3.7 Quantos dias você leva para preparar o seu lote para a queimada? ____

IV. Produção

4.1 As seguintes perguntas estão sobre suas colheitas.
        a. Quanto do seu lote você usou para _____ (cultivo) este ano?
        b. Ano passado? _____
        c. Quanto _____ (cultivo) você colheu no ano passado?
        d. Quanto _____ (cultivo) você vendeu no ano assado?
        e. A que custo?_____

Anuais (lavoura branca)	1. Área 1996	2. Área 1995	3. Unidade	4. Colheita 1995	5. Unidade	6. Vendeu S/N	7. Quan-to	8. Preço
a. Arroz								
b. Milho								
c. Feijão								
d. Mandioca								

4.2 a. Você usou fertilizantes químicos? Sim Não Em quais lavouras?
      b.Você usou fertilizantes orgânicos (como estrume)? Em quais?
      c. Você usou pesticidas? Sim Não Em quais lavouras?
      d. Você usou herbicidas? Sim Não Em quais lavouras?

	Arroz	Milho	Feijão	Mandioca	Café	Horta	
Fertilizantes Químicos							
Fertilizantes Orgânicos							
Pesticidas							
Herbicidas							

4.3 As seguintes perguntas estão sobre a sua colheita de arvores e frutas.

a. Quanto do seu lote você usou para _____ (cultivo) este ano?

b. Ano passado? _____

c. Quanto _____ (cultivo) você colheu no ano passado?

d. Quanto _____ (cultivo) você vendeu no ano passado?

e. A que custo? _____

Culturas Perenes Permanentes	1. Área 1996	2. Área 1995	3. Uni-dade	4. Colh-eita 1995	5. Uni-dade	6. Vend-eu S/N	7. Quan -to	8. Preço
a. Café								
b. Cacau								
c. Banana								
d. Cítricos								
e. Cupuaçú								
f. Pupunha								
g. Açaí								
h Acerola								
i. Mogno								
j. Madeira de lei								
k. Outras árvores								
l. Jaçu								
m. Abacati								
n. Manga								
o. Coco								
p. outra								

4.4 Onde você vendeu a sua colheita? _____

V. Gado e Pastagens

5.1 Sobre as suas cabeças do gado, quantas estão produzindo leite? _____

5.2 Alguém apanha o seu leite para vender no mercado?

5.3 Qual é importância do comércio de leite para sua renda?
    1. Muito importante
    2. Importante
    3. Pouco importante
    4. Não importante

5.4 Quanto de leite você produz à cada dia?
    5.4a na época de seca _____
    5.4b na época de chuva _____

5.5 Por quanto você vende o leite? _____

VI. Novas Técnicas de Cultivo

6.1 Você sabe quais são as novas técnicas de cultivo? Sim Não

Novas técnicas agrícolas podem contribuir para produção agrícola e ao mesmo tempo, conservar as florestas desta região. As novas técnicas incluem a combinação de culturas perenes com culturas anuais como: milho, arroz, feijão e mandioca. Esta lavoura usa a floresta, a qual não é brocada, derrubada nem queimada, em combinação com a lavoura como: consórcio e cultura casada. Além disso, você não depende totalmente da renda do gado (leite ou carne), mas de várias fontes de renda, como a maioria da renda derivada de culturas perenes ou da criação de abelhas e de peixe.

6.2 Você utiliza algumas das novas técnicas? Sim Não Se sim, quais? Se não, 6.8

6.3 a. Cultivando culturas perenes
    b. Deixando as árvores
    c. Consórcio (misturando as colheitas juntos) Qual colheita?
    d. Cultura casada (usando árvores com lavouras brancas)
       Quais árvores e quais lavouras brancas?
    e. Usando fontes alternativas de renda

6.4 Quando você começou a usar algumas destas técnicas?_____

6.5 Quanto do seu lote você usa para?
    a. Cultivar culturas perenes _____
    b. Deixar as árvores crescerem_____
    c. Consórcio _____
    d. Cultura casada _____
    e. Usando fontes alternativas de renda

6.6 Quais das seguintes fontes alternativos de renda você tem?
    a. criação de abelhas
    b. criação de peixe
    c. Outro (qual)

6.61a Quanto litros de mel colheu no ano passado? _____
6.61b Vendeu? _____
6.61c A que preço vendeu? _____

6.62a Quanto peixe colheu no ano passado? _____
6.62b Vendeu? _____
6.62c A que preço vendeu? _____

6.7 Qual é importância de _____ para a sua renda?
    1. Muito importante
    2. Importante
    3. Pouco importante
    4. Não importante
a. Criação de abelhas _____
b. Criação de peixes _____
c. Outro (qual) _____

6.8 As outras pessoas que moram nesta propriedade rural têm outras fontes de renda além da lavoura?

6.8a Quantas pessoas?_____

6.8b Quais são as fontes de renda? _____

6.9 (Se o agricultor usa novas técnicas) Por que você usa as novas técnicas?
  a. É mais proveitoso
  b. As semente foram fornecidas.
  c. Foi fornecido dinheiro.
  d. É melhor para as gerações futuras.
  e. É melhor para a terra e a floresta.
  f. Os meus pais usam esses métodos
  g. Os meus vizinhos usam esses métodos
  h. Outro (qual)

VII. O custo das novas técnicas (para os agricultores que as usam)

7.1a As novas técnicas requerem mais ou menos trabalhadores? Sim Não

7.1b Caso sim, quantos mais ou quantos menos?

7.2a As novas técnicas são mais caras do que as velhas técnicas? Sim Não

7.2b Caso sim, quanto?

7.3 Você usou a mesma área do seu lote para a lavoura depois de que começou à adotar as novas técnicas? Sim Não

7.4 Quanto colheu antes de adotar as novas técnicas (6.4) _____ anos passado?

Culturas	Colheita Antes	Unidade
a. Arroz		
b. Milho		
c. Feijão		
d. Mandioca		
e. Café		
f. Cacau		
g. Banana		
h. Cítricos		
i. Cupuaçú		
j. Pupunha		
k. Açaí		
l. Acerola		
m. Manga		
n. Coco		
o. Jaçu		
p. Abacati		
q. Madacusha		
r. outra		
s. outra		
t. outra		
u. outra		

VIII. Métodos Agrícolas

8.1 Estes itens tiveram algum efeito nos seus hábitos de cultivar?

1. muito efeito
2. algum efeito
3. nenhum efeito

a. Os seus vizinhos _____
b. Uma associação, um sindicato, ou uma cooperativa _____
c. Semente grátis _____
d. Árvores grátis _____
e. Empréstimos _____
f. Um mercado para vender as suas culturas anuais _____
g. Um mercado para vender as suas culturas perenes _____
h. As condições da terra (sabe sobre o solo de seu lote) _____
i. As árvores eram de alto valor para derrubar _____
j. A maneira de cultivo da terra usadas na região _____
k. A maneira de cultivo da terra usadas por seus pais _____

Surveyor's Comments:

Appendix D

Discrete Choice Models and Technology Adoption

D.1 DISCRETE CHOICE MODELS

Discrete choice models, also known as probability, multivariate or limited dependent variable models, are frequently used in the study of technology adoption in developing countries (Hussain *et al.* 1994, Adesina and Zinnah 1993, Lin 1991, Kebede *et al.* 1990, Shakya and Flinn 1985, Akinola and Young 1985). These include probit, logit, tobit and Heckman regression techniques. Discrete choice models are applicable in a wide variety of fields and are used extensively in survey or census-type data (Gujarati 1995). In the literature review of diffusion and adoption studies in Chapter 4 many of the papers used discrete choice analysis in order to determine the probability of adopting a new technology. The probability of adoption is used in these studies because it is an excellent indicator of adoption trends and therefore can indicate how the adoption of new technologies can be increased. For example, if the probability of adopting a new agricultural technology is low, the model may show that financial support or incentives can increase the adoption rate.

This appendix provides a review of discrete choice models used in the regression analysis in Chapter 6. In particular it discusses the tobit and Heckman techniques. These are discrete choice models that estimate first the probability of adoption and second the extent of adoption.

D.2 CHOOSING A PROBABILITY MODEL

Studies that use linear, log-linear or semi-logarithmic regression equations to

130

measure the adoption rate and the intensity of the use of an innovation often estimate the parameters using ordinary least squares (OLS). Such applications fail to capture the complexity of the decision-making process when truncated or censored data are used and produce misleading if not erroneous influences. Non-adopters for whom expenditure on the innovation is zero are excluded from the sample, creating a sample selection bias. This bias leads to a heteroskedastic error structure and inefficient regression parameters resulting in biased and inconsistent parameter estimates (Maddala 1983). The exclusion of non-adopters implies a zero demand. A more appropriate assumption – made in multivariate models such as probit, logit and Heckman estimations – is that non-adopters display market behavior by choosing not to adopt the technology. A comparison of these two groups could also serve to indicate any significant difference between the characteristics of the adopters and non-adopters.

Attempting to correct for the sample selection bias by including non-adopters in linear, log-linear and semi-logarithmic regression equations still results in a bias. As non-adopters, for whom the dependent variable is zero, are excluded from the sample, a sample selection bias is created and hence the regression coefficients are likely to be biased. The clustering of observations at zero, when correcting for the sample error, would violate the OLS assumption of a continuous dependent variable (Pindyck and Rubinfeld 1981). Feder *et al.* (1985) review the range of bivariate and multivariate statistical models used to analyze adoption behavior, criticize bivariate OLS models and conclude that they provide no insight into the relationship between adoption and the determinants of adoption. The authors recommend the use of probit or logit models to avoid most of these difficulties. Since the publication of that paper, there have been many studies that have used probit and logit models to study the adoption of new technologies in farming (Savagado *et al.* 1994, D' Souza *et al.* 1993, Kebede *et al.* 1990, Feder and Slade 1984).

Feder *et al.* (1985) review the adoption literature of developing countries and determine the best economic models that investigate the adoption decision to be probit and logit models. According to this review, the probit-logit methodology is preferable to discriminate analysis for analyzing the adoption decision since any form of linear regression results used to determine a discrete choice results in a bias. They also point out that few studies adopted procedures that explicitly account for the qualitative nature of the dependent variable and the intensity of adoption.

Since the Feder *et al.* (1985) study, multivariate statistical models have become common in the adoption decision literature with growing support for the tobit model (Tobin 1958). The tobit model measures not only the probability of adoption but also the intensity of use of the technology once it is adopted. Therefore, when the intensity of adoption is desired, the tobit model of estimation has been shown to be superior to both the logit and probit methods of estimation (Saha *et al.* 1994, Adesina and Zinnah 1993, Lin 1991, Gould *et al.* 1989, Norris and Batie 1987, Akinola and Young 1985, Shakya and Flinn 1985).

D.3 TOBIT MODEL SPECIFICATION

The tobit model (Tobin 1958) has been used by economists extensively since it was first developed in order to estimate the determinants of demand for a variety of consumer goods. Tobin measured the demand for luxury goods. These goods are unique in that they measure a zero demand in those years that consumers do not buy them. The zero demand becomes a problem when linear models of estimation are used. Instead, a censored probability model such as the tobit assumes that non-adopters are displaying market behavior by choosing not to purchase the good (Akinola and Young 1985). For example, the expenditures on cars may be used to derive a demand for cars, but the price that a consumer is willing to pay may be below the market price, and therefore the consumer will not purchase the car. The demand price would therefore be estimated to be zero, instead of the price below the market price. The OLS estimates of the parameter obtained from the subset of consumers (those who do purchase the car) will be biased as well as inconsistent as a result of this approach.[1]

The tobit model assumes that many variables have a lower (or upper) limit and take on this limiting value for a substantial number of respondents. For the remaining respondents, the variables take on a wide range of values above or below the limit. An explanatory variable in such a situation may influence both the probability of limit responses and the size of non-limit responses (Akinola and Young 1985).

The theoretical framework of the tobit model can be explained by the concept of a threshold value of the dependent variable. Given that the decision to adopt sustainable agriculture is characterized by a dichotomous choice between two mutually exclusive alternatives, (slash-and-burn agriculture and sustainable agriculture) there is a break-point in the dimension of explanatory variables such as farmer characteristics, capital assets, and information variables. There is a point below which a stimulus elicits no observable response. The independent variables which determine the extent of adoption of sustainable agriculture are only measured once the decision to use the new agricultural technique is made. Only when the strength of the stimulus exceeds that threshold level does adoption occur and the second decision on the intensity of its use is taken. In other words, an underlying continuous choice model on adoption exists but is only observed if sustainable agriculture is adopted.

For example, let Y, the dependent variable, denote the probability of adopting sustainable agriculture and X, a vector of explanatory variables, denote the variables that influence the probability of adoption. The Y variable can be defined in the context of a limited variable which takes on two values: $Y = Y^*$ if the decision results in the adoption of sustainable agriculture and $Y=0$ if adoption does not occur. The dichotomous nature and intensity of this decision is represented in Figure D.1. Point T represents the break-point or the threshold. At values of X greater than T there is a probability of 1 of adoption, and the intensity of adoption, represented by Y^* is continuous, shown by the line TC. At values of X below T, the probability of adoption is zero and the intensity of adoption is, of course, zero. In Figure D.1, the portion of the x-axis from 0 through to T represents the observation of those farmers who choose not to adopt sustainable agriculture, while the portion of the curve from T to C represents the extent of adoption for those farmers who do choose to use sustainable agriculture.

In the example just described, the dependent variable is censored.

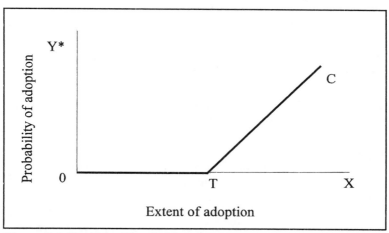

Figure D.1 The Tobit Threshold

Information is missing for the dependent variable when the choice is made to adopt slash-and-burn agriculture (not to adopt sustainable agriculture), pertaining to the extent of adoption. But the corresponding information for the independent variables is present[2] (Pindyck and Rubinfeld 1991).

The tobit model, like all limited dependent variable or discrete choice models, interprets the dependent variable as the probability of adoption and therefore limits the value of the dependent variable to be between 0 and 1. A transformation is used, therefore, to keep the values of the estimated dependent variables in the 0 to 1 range. The requirement of such a process is that it translates the values of the attributes X_i, which may range in value over the entire real line, into a probability which ranges in value from 0 to 1. It is desirable that the transformation should maintain the property that increases in X_i are consistently associated with increases or decreases in the dependent variable for all values of X_i, making a monotonic transformation necessary. The cumulative probability function (defined as the probability that an observed value of the variable X_i will be less than or greater than the threshold value) provides the necessary transformation for this operation. The range of the cumulative probability function is the interval (0,1) since all probabilities lie within these values. The standard cumulative normal distribution of $X_i\beta$ (the cumulative distribution function, CDF)[3] is given by:

$$F(X_i\beta)' = \int_{-\infty}^{X_i\beta'} \frac{1}{\sqrt{2\pi}} e^{-s^2/2} ds \qquad \text{D.1}$$

where s is a random variable that is normally distributed with mean zero and unit variance (Pindyck and Rubinfeld 1981). To estimate the parameters, β, a maximum likelihood method is applied.

The expected functional relationship of the adoption decision resembles a sigmoid curve for the tobit model. Since a probit estimation of the probability of adoption is first estimated, the sigmoid curve is based on the assumption that the cumulative distribution of the error term is normal (the logit model assumes that this distribution is logistic[4]). The main characteristic of a normal sigmoid curve is the existence of a lower limit (zero, for probability models) and an upper limit (1, for probability models) that bind the possible outcomes. A response in the decision variable can be observed only in the segment between the two extremes (Akinola and Young 1985). It may be assumed that the decisions of a population of potential adopters would follow an 'S'-shaped curve of this type, when the values of the explanatory variable exceed the threshold point. The 'S'-shaped curve follows the adoption diffusion theory discussed in Chapter 4. As Mansfield (1971) first showed when an innovation is introduced, a firm may be uncertain regarding its profitability and success, if it were to be adopted. This uncertainty can be reduced over time as the information about the innovation accumulates, and, as the information increases, the probability that the innovation will be adopted will also increase, resulting in an S-shape diffusion curve or a logistic (learning) curve. The rationale for this diffusion pattern through time is that the adoption of an innovation accelerates initially as it becomes more widely known and as the result of competitive pressure. Eventually, the rate of adoption slows as the number of firms not employing the innovation declines. The implication of this relationship is that when values of explanatory variables are close to the threshold point, policies that would increase the values of X are likely to encourage adoption.

The stochastic model of analysis of a simultaneous innovation decision making process of potential adopters may be expressed as follows:

$$Y_i = X_i\beta + \mu_i \quad if \quad Y^* = X_i\beta + \mu_i > T$$
$$Y_i = 0 \quad if \quad Y^* = X_i\beta + \mu_i \leq T \qquad i = 1,2,.........,n \qquad D.2$$

where n is the number of observations, Y_i is the limited dependent variable, Y_i^* is the continuous dependent variable when adoption occurs (in practice Y^* is unobservable, instead we observe the dummy variable, Y_i), X_i is a vector of explanatory variables, β is a vector of unknown coefficients, T is the threshold point, and μ_i is an independently distributed error term assumed to be normally distributed with a constant mean and constant variance, σ^2 (like the probit model).

Equation D.2 is a simultaneous and stochastic decision model. The tobit model estimates the probability, using probit estimation, and the intensity of adoption, under the maximum likelihood method, simultaneously. It is for this reason that the determinants both of the adoption and intensity decisions are assumed to be the same. When the unobserved variable Y^* is less than T, the observed qualitative variable Y_i that indexes adoption is zero. And when the unobserved variable Y^* is greater than T, the observed qualitative variable Y_i that indexes adoption is a continuous function of the explanatory variables.

In cases where substantially large numbers of decision-makers (at least more than 50 per cent) have completely adopted the new technology, a variant of the 1-limit tobit model in equation D.2 (such as the 2-limit tobit) could be used (Adesina and Zinnah 1993, Gould *et al.* 1989, Akinola and Young 1985).

D.3.1 Tobit Model Elasticities

In addition to measuring both the probability and intensity of adoption, the tobit model has the advantage that it gives information about the elasticity of the predicted probability. McDonald and Moffitt (1980) have shown that elasticities calculated at the means of the variables can be decomposed into two parts: the elasticity of the probability of being above the limit (elasticity of adoption) and the elasticity of the conditional expected value (elasticity of effort) given adoption occurs. The two elasticities add up to equal the total elasticity or the percentage change in the dependent variable given a 1 per cent change in the independent variable.

D.3.2 Tobit Estimation

Both the probit and tobit models use maximum likelihood methods to estimate the coefficients. The maximum likelihood method has a number of desirable statistical properties. The regression coefficients are asymptotically efficient, unbiased and consistent. In addition, all parameter estimators are asymptotically normal so that an analog of the t test is applied as a test for significance of the individual coefficients (Pindyck and Rubinfeld 1991). In the maximum likelihood model a log likelihood ratio test replaces the usual F test of the OLS regression models to evaluate the significance of all or a subset of coefficients. The goodness of fit of the model is inferred from an analogy of the coefficient of determination of OLS models. This pseudo-R^2 is calculated and interpreted in a similar manner as is R^2 in OLS regression models. However, the pseudo-R^2 should be used with caution as its sampling distribution is not known (Shakya and Flinn 1985).

The underlying tobit stochastic model may be expressed by the following relationship:

$$Y_i = X_i\beta + \mu_i \quad \textit{if} \quad Y^* = X_i\beta + \mu_i > 0$$
$$Y_i = 0 \quad \textit{if} \quad Y^* = X_i\beta + \mu_i \leq 0 \qquad i = 1,2,\ldots\ldots,n \quad \text{D.3}$$

which is the same relationship as in equation D.2 with a defined threshold value, T, of 0.

D.3.3 Influences of the Dependent Variable

The number of zero and non-zero responses can influence the impact that a change in the independent variables can make on the extent of adoption decision. In a study on the decision to purchase durable goods by Tobin (1958), 75 per cent of the observations had non-zero expenditures on durable goods. McDonald and Moffitt (1980) evaluated Tobin's results and found that 54 per cent of the total change in durable goods expenditure resulting from a change in the independent variables would be generated by marginal changes in the value of expenditures. (The value of expenditures is estimated secondly as the extent of adoption once the adoption decision is made). Therefore, McDonald and Moffitt determine that 46 per cent of the total change in durable goods expenditure resulting from a change in the independent variables would be generated by marginal changes in the probability of purchasing a durable good at all.

In the same paper, McDonald and Moffitt also evaluate data from a paper concerned with the demand for durable goods (Dagenais 1975). Dagenais' (1975) data contain fewer non-zero purchases. Only 21 per cent give non-zero responses. Evaluating at this point, McDonald and Moffitt find a lower percentage (23 per cent) of any total change that was due to marginal consumption changes. Most of the response, 87 per cent, is due to changes in the probability of purchasing the durable good. Therefore, the smaller (larger) the number of non-zero responses, the larger (smaller) the percentage of the total change in the extent of adoption decision will be generated by marginal changes in the extent decision, rather than the decision to adopt the new technology.

D.4 HECKMAN MODEL SPECIFICATION

The underlying assumption of the tobit model is that the same set of determinants which influence the adoption decision also influence the choice of the intensity of effort (Maddala 1983). The Heckman model (Heckman 1979) offers an alternative model to deal with censored samples, which allows for different variables that determine the adoption choice and then the effort allocated when adoption is chosen. The two-equation procedure involves the estimation of a probit model of the adoption decision, calculation of the sample selection control function and incorporation of that control function (the inverse Mills ratio) into the model of effort that is estimated with ordinary least squares (OLS). While the Heckman model does allow for different determinants between the adoption choice and the intensity of use, it does not allow for the decomposition of the elasticity afforded by the tobit procedure.

The Heckman model determines both the probability of adoption as well as the intensity of use, like the tobit model; however the assumption of the tobit model, that the same variables influence both decisions, is dropped. The two-equation procedure involves the estimation of a probit equation of adoption, and incorporation of that bias into a model of effort estimated using OLS.

The two-step procedure can be illustrated by the following equations:

$$Y_i = X_i\beta + \mu_i \quad if \quad Y^* = X_i\beta + \mu_i > 0$$
$$Y_i = 0 \quad\quad if \quad Y^* = X_i\beta + \mu_i \leq 0 \quad\quad i = 1,2,\ldots\ldots,n \quad D.4$$

Since the likelihood function for the probit model is well-behaved, the dummy variable is defined:

$$I_2 = 1 \quad\quad if \ Y_i > 0$$
$$I_2 = 0 \quad\quad otherwise \quad\quad D.5$$

Using the probit model, consistent variables of β/σ are calculated. Using these estimates, values of ϕ_i and Φ_i are calculated (where $\phi_i = \sigma f_i$ and $\Phi_i = F_i$; refer to equation D.1 and the coinciding endnote for further explanation). For each observation λ is calculated as:

$$\lambda = \phi(\alpha'/w) / \Phi(\alpha'w) \quad\quad D.6$$

Equation D.6 is known as the inverse Mills ratio. After calculating λ, equation D.6 is estimated using OLS (Maddala 1983).

The inverse Mills ratio, sometimes referred to as the hazard rate, is based on the probability density function of the censored error term. The cumulative distribution function is then derived from the probability density function. The

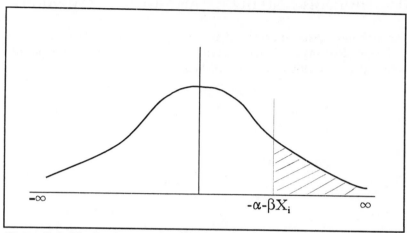

Figure D.2 The Probability Function of a Normal Random Variable

calculation of this ratio can best be explained in terms of Figure D.2.

The shaded area in Figure D.2 represents the cumulative density function. Dividing by this area normalizes the density function so that the total area under the probability density function is equal to one. The inverse Mills ratio can therefore be used to normalize the mean of the error terms to zero, and consistent estimators are then calculated for α and β, according to the normalization of the error terms. From the estimated parameters $\hat{\alpha}$ and $\hat{\beta}$ of the probit model, it is easy to calculate $\hat{\lambda}$. Because $\hat{\lambda}$ approaches λ as the sample size gets large, and λ normalizes the mean of the error terms, ordinary least squares estimation of equation D.7 yields consistent estimators of α and β:

$$Y_i = \alpha + \beta X_i + \sigma \hat{\lambda}_i + \mu_i \qquad\qquad \text{D.7}$$

In this equation $\hat{\lambda}$ is added as an explanatory variable. The two-stage estimator yields heteroskedastic errors (the error variance depends on X_i as well as whether $Y=0$, or not) so that the usual t-statistics are biased. In addition, the OLS method is not as efficient as the maximum likelihood method, and therefore the maximum likelihood method is preferred for this type of calculation.

D.5 THE TOBIT AND HECKMAN MODELS COMPARED

The tobit model can be explained as a special case of the Heckman model. Following Heckman's derivation (1979), consider a random sample of i observations. Equations for individual i are:

$$Y_{1i} = X_{1i}\beta_1 + U_{1i}$$
$$Y_{2i} = X_{2i}\beta_2 + U_{2i} \qquad i = 1, 2, \ldots\ldots, I \qquad\qquad \text{D.8}$$

$$E(U_{ji}) = 0 \quad E(U_{ji}, U_{j'i''}) = \sigma_{ij'} \quad i = i'' \qquad\qquad \text{D.9}$$
$$= 0 \quad\quad i \neq i''$$

where X_{ji} is a 1 x K_j vector of exogenous regressors, β_j is K_j x 1 vectors and the final assumption is a consequence of a random sampling scheme. The joint density of U_{1i}, U_{2i} is $h(U_{1i}, U_{2i})$. Suppose that data are missing on Y_1 for certain observations. (The Heckman model is used for censored samples.) The population regression equation, for Y_{1i} in equation D.7, can be written as:

$$E(Y_{1i} \mid X_{1i}) = X_{1i}\beta_1 \qquad\qquad \text{D.10}$$

If the conditional expectation of U_{1i} is zero, the regression function for the selected sub-sample is the same for as the general data. In the general case:

$$E(U_{1i} \mid X_{1i}) = E(U_{1i} \mid X_{1i}, Y_{2i} \geq 0)$$
$$= E(U_{1i} \mid X_{1i}, U_{2i} \geq X_{2i}\beta_2) \qquad\qquad \text{D.11}$$

The model outlined contains a variety of models as special cases, one of which is the tobit model. If $h(U_{1i}, U_{2i})$ is assumed to be singular and to have a normal density ($U_{1i} \equiv U_{2i}$) and $X_{2i} = X_{1i}$, $\beta_1 = \beta_2$, the tobit model takes shape. When U_{1i} and U_{2i} are independent the data on X_{1i} are missing randomly, the conditional mean of U_{1i} is zero. In the general case, it is non-zero and the sub-sample regression function is:

$$E(U_{1i}| X_{1i}, Y_{2i} \geq 0) = X_{1i}\beta_1 + E(U_{1i}| U_{2i} \geq X_{2i}\beta_2)$$ D.12

D.6 THE USE OF TOBIT AND HECKMAN MODELS IN THE STUDY

This study uses a combination of probit, tobit and Heckman techniques. Estimations of the probability of becoming an APA (Association of Alterative Producers) member, adopting sustainable agriculture and the extent to which sustainable agriculture is adopted are estimated using the techniques discussed in this chapter. These models as well as the data used in the analysis are presented in Chapter 5 and analyzed in Chapter 6.

ENDNOTES

1. The bias arises from the fact that if we only consider those observations from consumers who purchase the car, there is no guarantee that the expected value of the error term $E(\mu_i)$ is zero. And without $E(\mu_i) = 0$ there is no guarantee that the OLS estimates will be unbiased.
2. When both types of data are missing (the data responding both to the dependent and the independent variables), the dependent variable is considered to be truncated.
3. The standardized normal CDF can be determined as follows: The decision about whether to use sustainable agriculture or not by the *i*th farmer depends on the unobservable utility index, Y^* (which is determined by the explanatory variable X_i) in such a way that the larger the value of the index Y_i^*, the greater the probability of adoption. The index is expressed as $Y_i^* = X_i\beta$. For each farmer there is assumed to be a critical value, Y_i^{**}, such that if Y_i^* exceeds Y_i^{**}, the farmer will adopt sustainable agriculture. The threshold, Y_i^{**}, like Y_i^* is not observable, but if it is assumed to be normally distributed, with the same mean and variance as Y_i^*, it is possible to estimate the parameters of the index and to obtain information about the observable index. The probability that Y_i^{**} is less than or equal to Y_i^* can be computed as follows:

$$Y_i = Pr(Y=1) = Pr(Y_i^{**} \leq Y_i^*) = F(Y_i^*) = \frac{1}{\sqrt{2\pi}} \int_{-\infty}^{T_i} e^{-s^2/2}ds = \frac{1}{\sqrt{2\pi}} \int_{-\infty}^{\beta X_i} e^{-s^2/2}ds$$

which comes from the probability density function:

$$f(z) = \int\limits_{-\infty}^{\beta x_i} \frac{1}{\sqrt{2\pi}} \, e^{-\frac{1}{2}s^2}$$

where s is the variable X (a normally distributed variable with mean μ and variance σ^2) converted into a standardized normal variable, s, by the following transformation: $s = (x - \mu)$ / σ (Gujarati, 1995).

4. Since the cumulative normal distribution and the logistic distribution are very close to each other, with the exception of the tails, the two methods produce very similar results, with the exception of very large samples where there are many observations at the tails (Maddala 1983).

Bibliography

Abbott, M. and O. Ashenfelter (1976), 'Labour Supply, Commodity Demand, and the Allocation of Time', *Review of Economic Statistics*, 43, 389–411.

Adesina, A. A. and M. M. Zinnah (1993), 'Technology Characteristics, Farmers' Perceptions and Adoption Decisions: A Tobit Model Application in Sierra Leone', *Agricultural Economics*, 9, 297–311.

Adriance, J. (1995), 'Planting the Seeds of a New Agriculture: Living with the Land in Central America', *Grassroots Development*, 19 (1), 3–17.

Akinola, A. A. and T. Young (1985), 'An Application of the Tobit Model in the Analysis of Agricultural Innovation Adoption Processes: A Study of the Use of Cocoa Separating Chemical Among Nigerian Cocoa Farmers', *Oxford Agrarian Studies*, 14, 26–51.

Almeida, A. L. O. and J. S. Campari (1995), *Sustainable Settlement in the Brazilian Amazon*, New York: Oxford University Press.

Anderson, A. B. (1990), 'Deforestation in Amazonia: Dynamics, Causes, and Alternatives', in A. B. Anderson (ed.), *Alternatives to Deforestation: Steps Toward Sustainable Use of the Amazon Rain Forest*, New York: Columbia University Press.

Anderson, A. B. (1993), 'Deforestation in Amazonia: Dynamics, Causes, and Alternatives', in Simon Rietbergen (ed.), *The Earthscan Reader in Tropical Forestry*, London: Earthscan.

Anderson, A. B. and E. M. Ioris (1992), 'Valuing the Rain Forest: Economic Strategies by Small-Scale Forest Extractavists in the Amazon Estuary', *Human Ecology*, 20 (3), 337–369.

Argarwal, B. (1983), 'Diffusion of Rural Innovations: Some Analytical Issues and the Case of Wood-Burning Stoves', *World Development*, 11, 359–376.

Ashenfelter, O. and J. J. Heckman (1974), 'Estimation of Income and Substitution Effects in a Model of Family Labor Supply', *Econometrica* 42, 73–85.

Atkinson, A. B. and F. Bourguignon (1987), 'Income Distribution and Differences in Needs', in G. F. Feiwel (ed.), *Arrow and the Foundation of the Theory of Economic Policy*, London: Macmillian Press.

Baer, W. (1995), *The Brazilian Economy: Growth and Development*, Connecticut: Praeger.

Baker, D. C. and D. W. Norman (1990), 'The Farming Systems Research and Extension Approach to Small Farmer Development', in M. A. Alteiri and S. B. Hecht (eds), *Agroecology and Small Farm Development*, New York: CRC Press.

Balcer, Y. and S. P. Lippman (1984), 'Technology Expectations and the Adoption of Improved Technology', *Journal of Economic Theory*, 34, 292–318.

Battjees, J. (1988), 'A Survey of Secondary Vegetation in the Surrounding of Ararcuara, Amazonas, Columbia, Amsterdam', Hugo de Vries Laboratory, University of Amsterdam.

Bigsten, A. (1988), 'A Note on the Modelling of Circular Smallholder Migration', *Economic Letters*, 28, 87–91.

Binswanger, H. P. (1991), 'Brazilian Policies that Encourage Deforestation in the Amazon', *World Development*, 19 (7), 821–829.

Bognanno, M.F., J. S. Hixson and I. R. Jeffers (1974), 'The Short Run Supply of Nurses' Time', *Journal of Human Resources*, 9, 80–94.

Bourguignon, F. (1989), 'Family Size and Social Utility', *Journal of Econometrics*, 42, 67–80.

Browder, J. O. (1988), 'Public Policy and Deforestation in the Brazilian Amazon', in R. Repetto and M. Gills (eds), *Public Policies and the Misuse of Forest Resources*, Cambridge, MA: Cambridge University Press.

Browder, J. O. (March 1992), 'The Limits of Extractivism: Tropical Forest Strategies Beyond Extractive Resources', *BioScience*, 42 (3), 174–182.

Browder, J. O. (September 1994), 'Surviving Rondônia: The Dynamics of Colonist Farming Strategies in Brazil's Northwest Frontier', *Studies in Comparative International Development*, 29, 45–69.

Brown, K. and D. W. Pearce (1994), *The Causes of Deforestation*, London: University College London Press.

Bushbacher, R. J., C. Uhl and E. A. S. Serrão (1988), 'Abandoned Pastures in Eastern Amazonia II: Nutrient Stocks in the Soil and Vegetation', *Journal of Ecology*, 76, 682–699.

Camp, S. L. (1991), 'The Population, Environment, and Poverty Nexus', *Journal of Environmental Science and Health*, 12, 409–420.

Catsiapis, G. and C. Robinson (1982), 'Sample Selection Bias with Multiple Selection Rules: An Application to Student Aid Grants', *Journal of Econometrics*, 18, 351–368.

Cleaver, P. and S. Schreiber (June 1992), 'Reflections on Debt and the

Environment', *Finance and Development*, 12, 28–30.

Cockburn, A. (1988), 'Trees, Cows, and Cocaine: An Interview with Susanna Hecht', *Lua Nova Cultura e Politica*, 10, 34–55.

Cohn, M. S., S. A. Rae and R. I. Lerman (1970), 'A Micro Model of Labor Supply', BLS Staff Paper No. 4, US Department of Labor, Washington D. C.: Government Printing Office.

Corry, S. (July/August 1993), 'The Rainforest Harvest: Who Reaps the Benefit?', *The Ecologist*, 23 (4), 148–153.

Coxhead, I. A. and P. G. Warr (May 1991), 'Technical Change, Land Quality, and Income Distribution: A General Equilibrium Analysis', *American Journal of Agricultural Economics*, 73, 345–360.

D'Souza, G., D. Cypers and T. Phipps (October 1993), 'Factors Affecting the Adoption of Sustainable Agricultural Practices', *Agricultural and Resource Economics Review*, 18, 159–165.

Dagenais, N. (August 1975), 'Application of a Threshold Regression Model to Household Purchases of Automobiles', *The Review of Economics and Statistics*, 57 (2), 275–285.

Dale, V. H. and M. A. Pedlowski (Winter 1992), 'Farming the Forests', *Forum for Applied Research and Public Policy*, 7 (4), 20–21.

Dale, V. H., F. Southworth and M. Pedlowski (March 1994), 'Modeling Effects of Land Management in the Brazilian Amazonian Settlement of Rondônia', *Conservation Biology*, 8 (1), 196–206.

Dale, V. H., R. V. O'Neil, M. Pedlowski and F. Southworth (June 1993), 'Causes and Effects of Land-Use Change in Central Rondônia, Brazil', *Photogrammetric Engineering & Remote Sensing*, 59 (6), 997–1005.

Departamento de Estradas de Rodagem, Rondônia (DER) (1988), Map Title: 'Estado de Rondônia'. Porto Velho, Brasil.

Departamento de Estradas de Rodagem, Rondônia (DER) (1994), Map Title: 'Estado de Rondônia'. Porto Velho, Brasil.

Dixon, R. (1980), 'Hybrid Corn Revisited,' *Econometrica*, 48, 1451–1461.

Dourojeanni, M. J. (1988), 'Si el Arbol de la Quina Hablara', Lima, Fundacion Peruana para la Consevacion de la Naturaleza.

Editora Turística e Estatística LTDa (1994), 'Estado de Rondônia: Rodoviário, Político, e Estatístico: Nova Divisão Politíca', (Map) Gioãna, Goiás.

Ehui, S. K. and D. S. C. Spencer (1993), 'Measuring the Sustainability and Economic Viability of Tropical Farming Systems: A Model from Sub-Saharan Africa', *Agricultural Economics*, 9, 279–296.

Ehui, S. K., T. W. Hertel and P. V. Preckel (1990), 'Forest Resource Depletion, Soil Dynamics, and Agricultural Productivity in the Tropics', *Journal of Environmental Economics and Management*, 18, 136–154.

146   *Sustainable Agriculture in Brazil*

Fankhauser, S. (1993), 'Evaluating the Social Cost of Greenhouse Gas Emissions', unpublished paper, Center for Social and Economic Research on the Global Environment, University College London and University of East Anglia.

Fearnside, P. M. (1983), 'Deforestation Alternatives in the Brazilian Amazon: An Ecological Evaluation', *Interciencia*, 8 (2), 63–78.

Fearnside, P. M. (1990), 'Predominant Land Uses in Brazilian Amazonia', in A. B. Anderson (ed.), *Alternatives to Deforestation: Steps Toward Sustainable Use of the Amazon Rain Forest,* New York: Columbia University Press.

Fearnside, P. M. (December 1993), 'Deforestation in the Brazilian Amazon: The Effect of Population and Land Tenure', *Ambio*, 28 (8), 537–545.

Feder, G. and R. Slade (August 1984), 'The Acquisition of Information and the Adoption of New Technology', *American Journal of Agricultural Economics*, 66.

Feder, G., R. E. Just and D. Zilberman (1985), 'Adoption of Agricultural Innovations in Developing Countries: A Survey', *Economic Development and Cultural Change*, 33, 255–297.

Fink, A. and J. Kosecoff (1985), *How to Conduct Surveys: A Step-By-Step Guide*, California: Sage Publications.

Fowler, F. J. and T. W. Mangione (1990), *Standardized Survey Interviewing: Minimizing Interviewer Related Error*, Newbury Park, California: Sage Publications.

Fragasso, R. (June 1995), 'Micromanaging for Our Region's Future with Micro Loans', *Pittsburgh Business Times*, 14, 9–10.

Frohn, R. C., V. H. Dale and B. D. Jimenez (March 1990), *Colonization, Road Development and Deforestation in the Brazilian Amazon Basin of Rondônia*, Oak Ridge National Laboratory, Environmental Sciences Division, Publication No. 3394, ORNL/TM – 114770 Oak Ridge, TN.

Gafsi, S. and R. Terry (October 1979), 'Adoption of Unlike High-Yielding Wheat Varieties in Tunisia', *Economic Development and Cultural Change*, 28, 119–134.

Gentry, A. H. (1989), 'Northwest South America (Columbia, Ecuador, and Peru)', in D. G. Campbell and H. D. Hammond (eds), *Floristic Inventory of Tropical Countries*, New York: New York Botanical Gardens.

Goldenberg, J. (Summer 1992), 'Current Policies Aimed at Attaining a Model of Sustainable Development in Brazil', *Journal of Environment and Development*, 1 (1), 105–115.

Goodland, R. J. A. (Spring 1980), 'Environmental Ranking of Amazon Development Projects in Brazil', *Environmental Conservation*, 7 (1), 9–26.

Gould, B. W., W. E. Saupe and R. M. Klemme (May 1989), 'Conservation Tillage; The Role of Farm and Operator Characteristics and the Perception of Soil Erosion', *Land Economics*, 65 (2), 167–182.

Greene, B. A. (1973), 'Rate of Adoption of New Farm Practices in the Central Plains of Thailand', *Cornell International Bulletin*, (24), Ithaca, New York: Cornell University Press.

Greene, W. H. (1995), *LimDep Version 7.0 User's Manual*, Plainview, New York: Econometric Software, Inc.

Greene, W. H. (1997), *Econometric Analysis*, Upper Saddle River, New Jersey: Prentice Hall.

Griliches, Z. (1957), 'Hybrid Corn: An Exploration in the Economics of Technical Change', *Econometria*, 25 (4), 501–522.

Grimes, A., S. Loomis, P. Jahnige, M. Burnham, K. Onthank, R. Alarcon, W. P. Cuenca, C. C. Martinez, D. Neill, M. Balick, B. Bennett and R. Mendelson (November 1994), 'Valuing the Rain Forest: The Economic Value of Nontimber Forest Products in Ecuador', *Ambio*, 23 (7), 405–410.

Gujarati, D. N. (1995), *Basic Econometrics*, New York: McGraw Hill.

Guppy, N. (1984), 'Tropical Deforestation: A Global View', *Foreign Affairs*, 62 (4), 928–965.

Hall, E. T. (1976), *Beyond Culture*, New York: Anchor Press / Doubleday and Co.

Ham, J. C. (1982), 'Estimation of a Labour Supply Model with Censoring Due to Unemployment and Underemployment', *Review of Economic Studies*, 49, 335–354.

Hannan, T. H. and J. M. McDowell (Autumn 1984), 'The Determinants of Technology Adoption: The Case of the Banking Firm', *Rand Journal of Economics*, 15 (3), 328–335.

Harrison, D. A. (1983), *Behaving Brazilian: A Comparison of Brazilian and North American Behavior*, New York: Newbury House Publishers.

Hartshorn, G. S. (1992), 'Possible Effects of Global Warming on the Biological Diversity in Tropical Forests', in R. L. Peters and T. E. Lovejoy (eds), *Global Warming and Biological Diversity*, New Haven, US and London, UK: Yale University Press.

Hecht, S. B. (1981), 'Deforestation in the Amazon Basin: Magnitude, Dynamics and Soil Resource Efforts', *Studies in Third World Societies*, 13, 61–108.

Hecht, S. B. (1983), 'Cattle Ranching in the Eastern Amazon: Environmental and Social Implications', *The Dilemma of Amazonian Development*, Colorado: Westview Press.

Hecht, S. and A. Cockburn (1990), *The Fate of the Forest: Developers,*

Destroyers, and Defenders of the Amazon, London: The Penguin Group.

Heckman, J. J. (1976), 'The Common Structure of Statistical Models of Truncation, Sample Selection and Limited Dependent Variables and a Simpler Estimator for Such Models', *Annals of Economic and Social Measurement*, 5, 475–492.

Heckman, J. J. (January 1979), 'Sample Selection Bias as a Specification Error', *Econometria*, 47 (1), 153–161.

Holden, S. T. (1993), 'Peasant Household Modelling: Farming Systems Evolution and Sustainability in Northern Zambia', *Agricultural Economics*, 9, 241–267.

Holloway, M. (July 1993), 'Trends in Environmental Science: Sustaining the Amazon', *Scientific American*, 90–99.

Howells, C. (Fall 1993), 'Women's World Banking: An Interview with Nancy Barry', *The Columbian Journal of World Business*, 28, 20–32.

Hurther, A. P. and A. H. Rubenstein (1978), 'Market Penetration by New Innovations: The Technology Literature', *Technological Forecasting and Social Change*, 11, 197–221.

Hussain, S. S., D. Byerlee and P. W. Heisey (1994), 'Impacts of Training and Visit Extension System on Farmers' Knowledge and Adoption of Technology: Evidence from Pakistan', *Agricultural Economics*, 10, 39–47.

Instituto Nacional de Colonização e Reforma Agrária (1984), 'Município de Ouro Preto do Oeste' (Map) Ministério da Agricultura, Coordenadoria Especial do Estado de Rondônia.

Jaeger, W. K. and P. J. Matlon (February 1990), 'Utilization, Profitability, and the Adoption of Animal Draft Power in West Africa', *American Journal of Agricultural Economics*, 72, 35–48.

Jones, D. W., V. H. Dale, J. J. Beauchamp, M. A. Pedlowski and R. V. O'Neil (1995), 'Farming in Rondônia', *Resource and Energy Economics*, 17, 155–188.

Kahn, J. R. (1995), *The Economic Approach to the Environmental and Natural Resources*, Florida: The Dryden Press, Harcourt Brace College Publishers.

Kahn, J. R. and J. A. McDonald (1995), 'Third World Debt and Tropical Deforestation', *Ecological Economics*, 12, 107–123.

Kamien, M. T. and N. L. Schwartz (1982), *Market Structure and Innovation*, Cambridge: Cambridge University Press.

Kebede, Y., K. Gunjal and G. Coffin (1990), 'Adoption of New Technologies in Ethiopian Agriculture: The Case of Tegulet–Bulga District, Shoa Province', *Agriculture Economics*, 4, 27–43.

Killingsworth, M. R. (1983), *Labor Supply*, Cambridge: Cambridge University

Press.

Lena, P. (1981), 'Dinamica da Estrutura Agrária e o Aproveitamento dos Lotes Em Um Projecto de Colonização de Rondônia', Anais do Seminário 'Expansão da Fronteira Agropecuária e Meio Ambiente da América Latina', Brasilia 10 a 13 de Novembro de 1981. Universidade de Brasília, Departemento de Económia.

Lin, J. Y. (August 1991), 'Education and Innovation Adoption in Agriculture: Evidence from Hybrid Rice in China', *American Journal of Agricultural Economics*, 73, 713–723.

Luo, S. H. and C. R. Han (1990), 'Ecological Agriculture in China', in C. A. Edwards, R. Lal, P. Madden, R. H. Miller and G. House (eds), *Ecological Agriculture in China in Sustainable Agricultural Systems*, Florida: St. Lucie Press.

Lynne, G. D., J. S. Shonkwiler and L. R. Rola (February 1988), 'Attitudes and Farmer Conservation Behavior', *American Journal of Agricultural Economics*, 70 (1), 12–19.

Maddala, G. S. (1983), *Limited-Dependent and Qualitative Variables in Econometrics*, Cambridge: Cambridge University Press.

Mahar, D. (1989), *Government Policies and Deforestation in Brazil's Amazon Region*, Washington, D.C.: World Bank.

Mahar, D. and R. Schneider (1994), 'Incentives for Tropical Deforestation: Some Examples from Latin America', in K. Brown and D. W. Pearce (eds), *The Causes of Tropical Deforestation: The Economic and Statistical Analysis of Factors Giving Rise to The Loss of the Tropical Forests*, London: University College London Press.

Mansfield, M. (1971), *Technical Change*, New York, US: WW Norton.

McDonald, J. F. and R. A. Moffitt (May 1980), 'The Use of Tobit Analysis', *The Review of Economics and Statistics*, 107 (2), 318–321.

Miller, K. R., W. V. Reid and C. V. Barber (1993), 'Deforestation and Species Loss', in R. Dorfman and N. S. Dorfman (eds), *Economics of the Environment: Selected Readings*, New York, US and London, UK: W. W. Norton.

Miller, T. and G. Tolley (November 1989), 'Technology Adoption and Agricultural Price Policy', *American Journal of Agricultural Economics*, 71 (4), 847–857.

Moran, E. F. (1990), 'Private and Public Colonization Schemes in Amazonia', in D. Goodman and A. Hall (eds), *The Future of Amazonia: Destruction or Sustainable Development?*, New York: St Martins Press.

Moran, E. F. (1993), 'Deforestation and Land Use in the Brazilian Amazon', *Human Ecology*, 21 (1), 1–21.

150   *Sustainable Agriculture in Brazil*

Moran, E. F., E. Brondizio, P. Mausel and Y. Wu (May 1994), 'Integrating Amazonian Vegetation, Land Use, and Satellite Data', *BioScience*, 44 (5), 329–338.

Myers, N. (1984), *The Primary Source: Tropical Forests and Our Future*, New York: W. W. Norton.

Myers, N. (1994), 'Tropical Deforestation: Rates and Patterns', in K. Brown and D. W. Pearce (eds), *The Causes of Tropical Deforestation: The Economic and Statistical Analysis of Factors Giving Rise to The Loss of the Tropical Forests*, London: University College London Press.

Nair, P. K. R. (1992), 'Agroforestry Systems Design: An Ecozone Approach', in N. P. Sharma (ed.), *Managing the World's Forests*, Iowa: Kendal/Hunt Publishing.

National Academy Press (1992), *Conserving Biodiversity: A Research Agenda for Development Agencies*, Washington, D.C.: National Academy Press.

Nepstad, D., C. Uhl and E. A. Serrão (1990), 'Surmounting Barriers to Forest Regeneration in Abandoned, Highly Degraded Pastures: A Case Study from Paragominas, Pará, Brazil', in A. B. Anderson (ed.), *Alternatives to Deforestation: Steps Toward Sustainable Use of the Amazon Rain Forest*, New York: Columbia University Press.

Norris, P. E. and S. S. Batie (July 1987), 'Virginia Farmers' Soil Conservation Decisions: An Application of Tobit Analysis', *Southern Journal of Agricultural Economics*, 19 (1), 79–90.

Nowak, P. J. (Summer 1987), 'The Adoption of Agricultural Conservation Technologies: Economic and Diffusion Explanations', *Rural Sociology*, 52 (2), 209–220.

Onis, J. D. (1992), *The Green Cathedral: Sustainable Development of the Amazon*, Oxford, New York: Oxford University Press.

Opare, K. D. (Spring 1977), 'The Role of Agricultural Extension in the Adoption of Innovations by Cocoa Growers in Ghana', *Rural Sociology*, 42 (1), 72–82.

Oster, S. (Spring 1982), 'The Diffusion of Innovation Among Steel Firms: The Basic Oxygen Furnace', *The Bell Journal of Economics*, 13 (1), 45–56.

Pampal, F. Jr. and J. C. Van Es (Spring 1977), 'Environmental Quality and Issues of Adoption Research', *Rural Sociology*, 42 (1), 57–71.

Panayotou, T. and S. Sungsuwan (1994), 'An Econometric Analysis of the Causes of Tropical Deforestation: The Case of Northeast Thailand', in K. Brown and D. W. Pearce (eds), *The Causes of Tropical Deforestation: The Economic and Statistical Analysis of Factors Giving Rise to The Loss of the Tropical Forests*, London: University College London Press.

Pearce, D. and K. Brown (1994), 'Saving the World's Tropical Forests', in K.

Brown and D. W. Pearce (eds), *The Causes of Tropical Deforestation: The Economic and Statistical Analysis of Factors Giving Rise to The Loss of the Tropical Forests*, London: University College London Press Limited.

Pearce, F. (September 1989), 'Kill or Cure? Remedies for the Rainforest', *New Scientist*, 16, 40–43.

Pedlowski, M. A. and V. H. Dale (September 1992), 'Land Use Practices in Ouro Preto do Oeste, Rondônia, Brazil', Oak Ridge National Laboratory, Environmental Sciences Division, Publication No. 3850, ORNL/TM – 114770, Oak Ridge, TN.

Pedlowski, M. A. and V. H. Dale, E.A.T Matricardi, and E. P. da Silva Filho, (1997), 'Patterns and Impacts of Deforestation in Rondônia, Brazil', *Landscape and Urban Planning* 38, 149–157.

Pindyck, R. S. and D. L. Rubinfeld (1981), *Econometric Models and Economic Forecasts*, New York: McGraw-Hill.

Pindyck, R. S. and D. L. Rubinfeld (1991), *Econometric Models and Economic Forecasts*, New York: McGraw-Hill.

Posel, S. and J. C. Ryan (March–April 1991), 'Toward Sustainable Forestry Worldwide', *Journal of Soil and Water Conservation*, 3, 119–122.

Price, M. (1994), 'Ecopolitics and Environmental Nongovernmental Organizations in Latin America', *Geographical Review*, 84, 42–60.

Prinsley, R. T. (1992), 'The Role of Trees in Sustainable Agriculture – An Overview, *Agroforestry Systems*, 20, 87–115.

Rahm, M. R. and W. E. Huffman (November 1984), 'The Adoption of Reduced Tillage: The Role of Human Capital and Other Variables', *American Journal of Agricultural Economics*, 66, 405–413.

Ransom, M. R. (1987), 'The Labor Supply of Married Men: A Switching Regressions Model', *Journal of Labor Economics*, 5 (1), 63–75.

Renkow, M. (February 1993), 'Differential Technology Adoption and Income Distribution in Pakistan: Implications for Research Resource Allocation', *American Journal of Agricultural Economics*, 75, 33–43.

Rodrigues, J. H. (1967), *The Brazilians: Their Character and Aspirations*, Austin, Texas: University of Texas Press.

Ruddell, E. (1995), 'Growing Food For Thought: A New Model of Site-Specific Research from Bolivia', *Grassroots Development*, 19 (1), 3–17.

Saha, A., H. A. Love and R. Schwart (November 1994), 'Adoption of Emerging Technologies Under Output Uncertainty', *American Journal of Agricultural Economics*, 76, 836–846.

Sands, D. M. (1986), 'The Technology Applications Gap: Overcoming Constraints to Small Farm Development', Food and Agricultural Organization of the United Nations, Rome.

152   *Sustainable Agriculture in Brazil*

Santos, B. A. (1983), *Amazônia: Potencial Mineral e Perspectivas de Desenvolovimento*, T. A. Querióz (ed.), São Paulo.
Savagado, K., T. Reardon and K. Pietola (August 1994), 'Farm Productivity in Burkina Faso: Effects of Animal Traction and Nonfarm Income', *American Journal of Agricultural Economics*, 76, 608–617.
Schneider, R. (1992a), 'Global Climate Change: Ecological Effects', *Interdisciplinary Science Review*, 17 (2), 142–148.
Schneider, R. (1992b), 'Brazil: An Analysis of Environmental Problems in the Amazon', Report 9104–BR, World Bank.
Schultz, T. W. (1964), *Transforming Traditional Agriculture*, New Haven, Connecticut: Yale University Press.
Secretaria de Estado de Desenvolvimento Ambiental (SEDAM) (1996), 'Mapa Politico e Adminstrativo de Rondônia'.
Serrão, E. A. and J. M. Toledo (1990), 'The Search for Sustainability in Amazon Pastures', in A. B. Anderson (ed.), *Alternatives to Deforestation: Steps Toward Sustainable Use of the Amazon Rain Forest*, New York: Columbia University Press.
Shakya, P. B. and J. C. Flinn (September 1985), 'Adoption of Modern Varieties and Fertilizer Use on Rice in the Eastern Tarai of Nepal', *Journal of Agricultural Economics*, 36 (3), 409–419.
Shishko, R. and B. Rostker (June 1976), 'The Economics of Multiple Job Holding', *American Economic Review*, 66, 298–308.
Silberberg, E. (1990), *The Structure of Economics: A Mathematical Analysis*, New York: McGraw Hill.
Skole, D. L., W. H. Chomentowski, W. A. Salas and A. D. Nobre (May 1994), 'Physical and Human Dimensions of Deforestation in Amazonia', *BioScience*, 44 (5), 314–321.
Southgate, D. (1992), 'Policies Contributing to Agriculture Colonization of Latin America's Tropical Forests', in N. P. Sharma (ed.), *Managing the World's Forests*, Iowa: Kendal/Hunt Publishing.
Southgate, D. (1994), 'Tropical Deforestation and Agricultural Development in Latin America', in K. Brown and D. W. Pearce (eds), *The Causes of Tropical Deforestation: The Economic and Statistical Analysis of Factors Giving Rise to The Loss of the Tropical Forests*, London: University College London Press.
Stackhouse, C. (July 1994), 'A Tale of Two Sisters', *International Wildlife*, 24, 16–23.
Strauss, J., M. Barbosa, S. Teixeira, D. Thomas and R. Gomes, Jr. (1991), 'Role of Education and Extension in the Adoption of Technology: A Study of Upland Rice and Soybean Farmers in Central-West Brazil', *Agricultural*

Economics, 5, 341–359.

Tietenberg, T. (1988), *Environmental and Natural Resource Economics*, Glenview, Illinois: Scott Foresman.

Tobin, J. (1958), 'Estimation of Relationships for Limited Dependent Variables', *Econometria*, 26, 24–36.

Tunali, I. (1986), 'A General Structure for Models of Double-selection and an Application to Joint Migration/Earnings Process with Remigration', in R. G. Ehrenberg (ed.), *Research in Labor Economics*, 8 (B), JAI Press.

Tunali, I., J. R. Behrman and B. L. Wolfe (June 1980), 'Identification, Estimation and Prediction Under Double Selection', Paper presented at the Joint Meetings of ASA and Biometric Society, Houston, Texas.

Uhl, C., R. Buschbacher and A. Serrão (1988), 'Abandoned Pastures in Eastern Amazonia. I: Patterns of Plant Succession', *Journal of Ecology*, 76, 663–681.

Uhl, C., D. Nepstad, R. Buschbacher, K. Clark, B. Kauffman and S. Subler (November/December 1989), 'Disturbance and Regeneration in Amazonia: Lessons for Sustainable Land Use', *The Ecologist*, 19 (6), 235–240.

US Department of Commerce, National Trade Bank and Economic Bulletin Board (1997), 'Brazil: Domestic Economy: Country Commercial Guides', http://www.stat–usa.gov/.

Varian, H. R. (1992), *Microeconomic Analysis*, New York: W. W. Norton.

Villacchia, H., J. E. Silva and C. M. C. da Rocha (1990), 'Sustainable Agricultural Systems in the Humid Topics of South America', in C. A. Edwards, R. Lal, P. Madden, R. H. Miller and G. House (eds), *Ecological Agriculture in China in Sustainable Agricultural Systems*, Florida: St. Lucie Press.

Waxler, T. (January 1994), 'Working Capital: A High Performance Third World Solution to Enterprise Development in the United States', *Children Today*, 23, 6–8.

Whitcover, J. and S. A. Vosti, (April 1996), 'A Socio Economic Characterization Questionnaire for the Brazilian Amazon: A Description and Discussion of Questionnaire Application Issues (Fieldwork 29 August–11 September 1994)', MP–8 Working Paper No. US96–001, A Research Program of the Environmental and Production Technology Division of the International Food Policy Research Institute.

Whitcover, J. and S. A. Vosti, (June 1996), 'Alternatives to Slash-and-Burn Agriculture (ASB): A Characterization of Brazilian Benchmark Sites of Pedro Peixoto and Theobroma (August/September 1994)', MP–8 Working Paper No. US96–003, A Research Program of the Environmental and Production Technology Division of the International Food Policy Research Institute.

Index

Sustainable Agriculture in Brazil

59-61
in Rondônia 18
see also slash-and-burn agriculture;
    sustainable agriculture
value 15, 27, 28, 55, 58, 106, 107
see also agricultural lots
Latin America
sustainable agriculture in 21-2
see also Amazon; Brazil
Lin, J.Y. 36, 37, 69, 78, 101, 130, 132
loans 22, 23
see also financial support; national
    debt
local concerns, on deforestation 6, 8
local market failures 27-8, 30-31
logging, statistics 11
logit models 131
lots, agricultural see agricultural lots
Luo, S.H. 19
Lynne, G.D. 36, 40

McDonald, J.A. 27, 28
McDonald, J.F. 136, 137
McDowell, J.M. 32
Maddala, G.S. 131, 137, 138
Mahar, D. 1, 2, 7, 12, 14, 17, 104, 105,
    106
Mangione, T.W. 45
Mansfield, M. 33-4, 135
marginal costs, of deforestation 29
markets
development of 27-8, 31
failure of 61
and deforestation 15, 25-31
Matlon, P.J. 23, 36, 39, 101
micro-loans 22, 23
migrants see settlers
Miller, K.R. 7, 28
Miller, T. 23, 36, 39, 101
mining 11, 13
models
of innovation adoption 33-4, 35-43,
    130-41
in sustainable agriculture 66-70
of survey analysis 70-73
Moffitt, R.A. 136, 137
Moran, E.F. 2, 11, 16, 27, 103, 106
multivariate statistical models 131-2

see also Heckman model; tobit model
Myers, N. 1, 8, 9, 10, 11, 12, 17, 23

Nair, P.K.R. 18, 19, 20
national concerns, on deforestation 6, 8
national debt, Brazil 27
National Integration Program 104
national market failures 27
neighboring effects, and extent of
    intercropping 79
Nepstad, D. 9
new technologies see innovations
non-timber forest products 13
markets for 28, 31
value 27-8
see also honey; rubber
Norris, P.E. 23, 36, 39, 69, 72, 101, 132
Nowak, P.J. 35
nutrients, in rain forest soil 8-9, 18

Opare, K.D. 2, 31, 36, 38, 65, 100
Operation Amazonia 8, 12-13, 14, 23,
    27, 103-7
ordinary least squares, disadvantages
    131
Oster, S. 32
Ouro Preto do Oeste
agricultural choices 33, 64, 65-6, 71
agricultural lots see agricultural lots
description of 2-3, 12-14, 46-7
farmers, characteristics see farmers,
    characteristics
harvests see harvests
land use 55, 57, 58, 59-61
land value 27, 55, 58
map 47
population 46-7, 48-9
sustainable agriculture in
    see sustainable agriculture,
    in Ouro Preto do Oeste
outreach programs
on sustainable agriculture 21
see also education programs;
    information
over-population, Brazil 8, 27, 30, 105

Panayotou, T. 15
pasture see cattle ranching